Creep in Timber Structures

RILEM REPORTS

RILEM Reports are prepared by international technical committees set up by RILEM. The International Union of Testing and Research Laboratories for Materials and Structures. Further information about RILEM is given at the back of the book.

1   Soiling and Cleaning of Building Facades
     Report of TC 62–SCF, edited by L.G.W. Verhoef

2   Corrosion of Steel in Concrete
     Report of TC 60–CSC, edited by P. Schiessl

3   Fracture Mechanics of Concrete Structures: from Theory to
     Applications
     Report of TC 90–FMA, edited by L. Elfgren

4   Geomembranes – Identification and Performance Testing
     Report of TC 103–MGH, edited by A. Rollin and J.M. Rigo

5   Fracture Mechanics Test Methods for Concrete
     Report of TC 89–FMT, edited by S.P. Shah and A. Carpinteri

6   Recycling of Demolished Concrete and Masonry
     Report of TC 37–DRC, edited by T.C. Hansen

7   Fly Ash in Concrete: Properties and Performance
     Report of TC 67–FAB, edited by K. Wesche

8   Creep in Timber Structures
     Report of TC 112–TSC, edited by P. Morlier

9   Disaster Planning, Structural Assessment, Demolition and
     Recycling
     Report of Task Force 2 of TC 121–DRG, edited by C. De Pauw
     and E.K. Lauritzen

10   Application of Admixtures in Concrete
     Report of TC 84–AAC, edited by A.M. Paillere

11   Interfaces in Cementitious Composites
     Report of TC 108–ICC, edited by J.-C. Maso

Creep in Timber Structures

Report of RILEM Technical Committee 112–TSC

RILEM
(The International Union of Testing and Research Laboratories
for Materials and Structures)

Edited by

P. Morlier
Laboratoire de Rhéologie du Bois de Bordeaux (LRBB),
Bordeaux, France

CRC Press
Taylor & Francis Group
Boca Raton London New York

CRC Press is an imprint of the
Taylor & Francis Group, an **informa** business

A TAYLOR & FRANCIS BOOK

CRC Press
Taylor & Francis Group
6000 Broken Sound Parkway NW, Suite 300
Boca Raton, FL 33487-2742

First issued in paperback 2019

© 1994 RILEM
CRC Press is an imprint of Taylor & Francis Group, an Informa business

No claim to original U.S. Government works

ISBN-13: 978-0-419-18830-8 (hbk)
ISBN-13: 978-0-367-44927-8 (pbk)

Typeset in 10.5/12 pt Times by Florencetype Ltd, Stoodleigh, Devon

A catalogue record for this book is available from the British Library

Publisher's Note
The publisher has gone to great lengths to ensure the quality of this reprint
but points out that some imperfections in the original may be apparent

Visit the Taylor & Francis Web site at
http://www.taylorandfrancis.com

and the CRC Press Web site at
http://www.crcpress.com

Contents

Contributors – RILEM Technical Committee 112–TSC

D.G. Hunt
Mechanical Engineering, Design and Manufacture Department
South Bank University
103 Borough Road, London SE1 0AA, UK

C. Le Govic
Laboratoire de Mécanique–Rhéologie
Centre Technique Bois et Ameublement (CTBA)
10 avenue de Saint-Mandé, 75012 Paris, France

P. Morlier
(Chairman)
Laboratoire de Rhéologie du Bois de Bordeaux (LRBB)
BP 10, 33610 Cestas-Gazinet, France

L.C. Palka
Forintek Canada Corporation
2665 East Mall, Vancouver, B.C. V6T 1W5, Canada

A. Ranta-Maunus
Laboratory of Structural Engineering
VTT (Technical Research Centre of Finland)
P.O. Box 26, Kemistintie 3, SF - 02151 Espoo, Finland

S. Thelandersson
Department of Structural Engineering
Lund Institute of Technology
P.O. Box 118, S - 221 00 Lund, Sweden

T.A.C.M. Van der Put
Stevin Laboratory
P.O. Box 5048
NL 2600 GA Delft, The Netherlands

Preface

RILEM Technical Committee 112 held its first meeting in Paris in May 1989. During the first sessions of this committee a general objective was defined: to collect maximum information about creep of wood, timber and timber structures, with particular attention to the interaction of moisture changes with mechanical loading which may lead to excessive deflections in timber structures. Indeed it is a quite modern scientific field to describe carefully, to try to explain, and to model correctly these phenomena for application to wooden structures (or timber drying).

While we were working on this state-of-the-art report, young scientists, in our laboratories, were progressing very efficiently so that this report is undoubtedly out of date for them. On the other hand, we apologize for delivering a non-exhaustive survey: some interesting contributions have certainly been neglected and we stopped our survey with papers published in 1990.

Thanks to occasional participants to this Technical Committee: U. Korin, S. Mohager, M. Piazza, T. Van der Put (who contributed Chapter 4 of this book).

Thanks to three members of my staff who greatly helped me to finalize this book: F. Generale, C. Lavergne, J. Laurent.

Thanks to my co-authors for the quality of their homework and for the friendly climate of our working sessions.

Pierre Morlier, Chairman

1

The importance of deflection requirements in the design of timber structures

S. Thelandersson

1.1 Introduction

According to the present draft of Eurocode 5 (December 1991) the design load combination in the serviceability limit state is defined as:

$$\sum G_{k,j} + Q_{k,1} + \sum_{l \triangleright 1} \vartheta_{1,i} Q_{k,i}$$

(1.1)

where $G_{k,j}$ are permanent loads

and $Q_{k,i}$ are variable loads

The subscript k denotes characteristic values and $\vartheta_{1,i} Q_{k,i}$ is the so-called frequent value of variable load i.

As far as creep deflections are concerned the final deflection u_{fin} will be calculated for each load type as:

$$u_{fin} = u_{inst}(1 + k_{def})$$

(1.2)

where u_{inst} is the instantaneous deflection of the load considered and k_{def} is a 'creep' factor depending on load duration class and service class. For load combinations, c.f. Eq.(1.1), the deflections will be calculated for each load separately and then added.

Recommended limits of deflection are given in Eurocode 5. For long-term deflections the following limits are proposed:

$$u_{2,max} \leq L/300$$

(1.3a)

$$u_{nett,max} \leq L/200$$

(1.3b)

where u_2 and u_{nett} are defined in Fig. 1.1.

The importance of an accurate prediction of deflections in the serviceability limit state depends to what extent deflection criteria are decisive for the design of typical structural elements. To illustrate the practical significance of deflection requirements in comparison with ultimate limit state design some results are given in what follows for straight beams made of glulam and timbers. Unless otherwise stated the

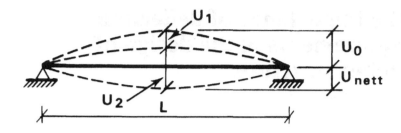

Fig. 1.1 Deflection of a structural beam. Definition of different components of deflection.
U_0 = precamber (if applied)
U_1 = deflection due to permanent loads
U_2 = deflection due to variable loads

calculations are based on the rules and numbers given in the Eurocode 5 draft from December 1991.

It is assumed in all cases that the beam is loaded with only one variable load Q_k (e.g. imposed load or snow load) in addition to the permanent load g_k. Only rectangular beams with width b and depth h are considered. The beams are assumed to be rectangular, spaced with spacing c, and the loads g_k and Q_k are uniformly distributed and specified per unit area.

1.2 Glulam beams

To compare the ultimate and serviceability limit states, the required beam depth h_s given by a serviceability criterion and the required beam depth h_u given by the ultimate limit state were calculated for different cases. The ratio between h_s and h_u is taken as an indicator of which of the criteria is decisive in the design of the beam. For h_s/h_u greater than 1 the dimensions of the beam are governed by the deflection limit.

Figure 1.2 displays the ratio h_s/h_u as a function of the load parameter $(Q_k + G_k)$ c/b. Calculations were made with each of the two criteria in Eq. (1.3a) and (1.3b) respectively. The results in Fig.1.2 are valid for glulam GK 37 with $f_{m,k}$ = 37MPa and $E_{o,mean}$ = 14 500. The ratio $G_k/(G_k + Q_k)$ is 0.25 in Fig. 1.2(a) and 0.5 in Fig 1.2(b). The variable load Q_k belongs to the load duration class 'medium term'.

The beam dimension is determined by the deflection requirement for low values of the load parameter $(G_k + Q_k)$ c/b and by the failure criterion for high values of this parameter. The deflection criteria (1.3a) and (1.3b) give almost the same results when the permanent loads are low compared to the total load (Fig. 1.2a). For higher ratios of $G_k/(G_k + Q_k)$ criterion (1.3b) applies (Fig. 1.2b).

If the variable load is assumed to be of the load duration class 'long term' a similar picture is obtained, but the values h_s/h_u become slightly smaller: 3.2% and 4.5% for $G_k/(G_k + Q_k) = 0.25$ and 0.5 respectively.

The practical range of the parameter $(G_k + Q_k)$ c/b for glulam is between 20 and 150 kN/m². Table 1.1 shows the values of this parameter for realistic combinations of c and b for glulam applications and for two different values of $G_k + Q_k$. The value $G_k + Q_k = 2.5$ kN/m² is representative for floors with imposed loads and roofs in regions with high snow loads whereas the value 1.25 kN/m² is representative for roofs with low snow loads.

It is quite evident that for high quality glulam, the deflection criteria are decisive in a majority of cases in practice.

Table 1.1 Values of the parameter $(G_k + Q_k)$ c/b in kN/m² for typical use of glulam beams

$G_k + Q_k$ (kN/m²)	b (mm)	c (m)			
		1.5	3.0	4.5	6.0
1.25	56	33	67	–	–
	90	21	42	63	–
	115	16	33	49	65
	165	–	23	34	45
	215	–	–	26	35
2.5	56	67	134	–	–
	90	42	83	125	–
	115	33	65	98	130
	165	–	45	68	91
	215	–	–	52	70

1.3 Timber beams

The ratio h_s/h_u is given in Fig. 1.3 as a function of the load parameter $(G_k + Q_k)$ c/b for timber beams of different strength classes. For higher strength classes, the deflection criterion is decisive especially for low values of the load parameter. The results in Fig. 1.3 are valid for the cases when the variable load Q_k belongs to the load duration class 'medium term'. For the case of 'long term' variable load the value of h_s/h_u becomes 3.2% lower.

For timber beams the practical range of the parameter $(G_k + Q_k)$ c/b is between 10 and 50. A very frequent case in the practical use of timber beams is a floor with $G_k \approx 0.6$ kN/m², $Q_k = 2$ kN/m², c = 600mm and b = 45 mm. In this case the parameter $(G_k + Q_k)$ c/b ≈ 35 kN/m².

4   DEFLECTION IN TIMBER STRUCTURES

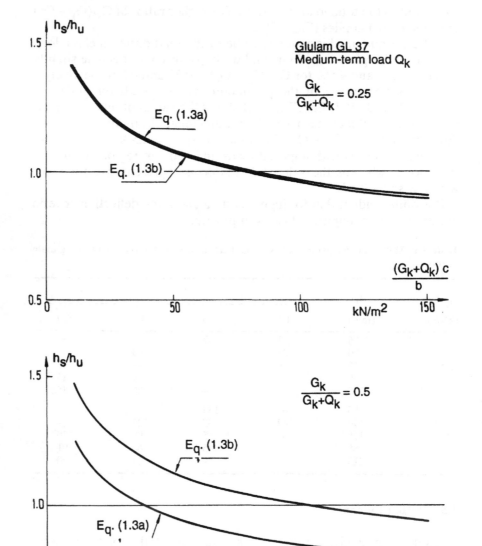

Fig. 1.2 The ratio between the required beam depths h_s and h_u in the serviceability and ultimate limit state respectively.

For this special case the ratio h_s/h_u is given in Fig. 1.4 as a function of the choice of deflection limit for $u_{nett, max,}$ cf. Eq (1.3b). The results in Fig. 1.4 are valid for the case when the variable load Q_k is classified as 'long term'. Fig. 1.4 shows that more severe deflection criteria than that recommended in Eurocode 5 lead to values of h_s/h_u larger than 1 for all strength classes.

$$G_k/(G_k + Q_k) = 0.25; (G_k + Q_k)\ c/b + 35\ kN/m^2$$
$$Q_k = \text{'long term' load}$$

As an aid to interpreting the results in Fig. 1.3, values of $(G_k + Q_k)\ c/b$ for realistic timber beam applications are given in Table 1.2.

Table 1.2 Values of the parameter $(G_k + Q_k)\ c/b$ in kN/m^2 for typical use of timber beams

$(G_k + Q_k)$	b	c (mm)			
(kN/m^2)	(mm)	300	450	600	900
	45	8	13	17	25
1.25	70	5	8	11	16
	95	8	6	8	12
	45	17	25	33	50
2.5	70	11	16	21	32
	95	8	12	16	24

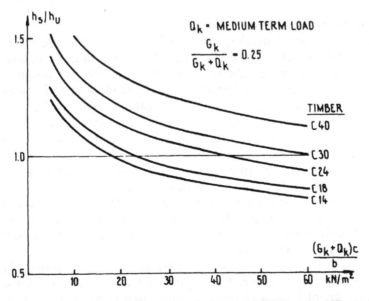

Fig. 1.3 The ratio between required beam depths h_s and h_u in the serviceability and ultimate limit state respectively. Deflection criterion according to Eq. (1.3b)

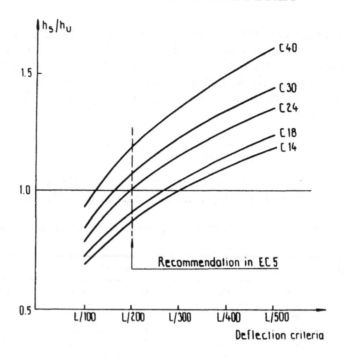

Fig. 1.4 The ratio h_s/h_u as a function of the chosen deflection limit for $u_{nett,max}$.

1.4 Sensitivity studies

As shown above, deflection criteria will often be decisive in practical design. It is therefore very important that both short-term and long-term deflections are predicted in an accurate way in design. Calculations of deflections are based on elastic moduli (mean values) and deformation factors k_{def}. It could be of interest to study the sensitivity of these factors to the design of timber and glulam beams. Figure 1.5 shows the relative change in h_s associated with a change in elastic modulus.

All changes are referred to the values obtained from the current version of Eurocode 5. The results in Figure 1.5 are valid for all strength classes of timber and glulam. For instance, if the elastic modulus for a certain grade is decreased by 20% relative to the value given in the current code version the beam depth required to fulfil the deflection limit is increased by about 8%.

The sensitivity to the choice of creep factors k_{def} is show in Fig. 1.6. It has been assumed here that k_{def} for all load duration classes including permanent load is changed by the same percentage. It is evident that the required beam depth is quite insensitive to the choice of creep

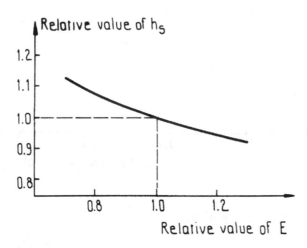

Fig. 1.5 Sensitivity of h_s to change in elastic modulus.

factors. For instance if all creep factors are doubled, h_s will increase only by 8–12%. This result is also insensitive to the ratio between permanent load G_k and the total load $G_k + Q_k$.

Conclusion

It was found that limits on deflection such as those recommended in the current version of Eurocode 5 are often very decisive in practical design, provided that there is a need to limit deflection. For long-span glulam beams limitation of deflections could be met by providing a precamber of the beam. This possibility has not been considered in the present analysis. However, it is clear that an accurate prediction of long-term as well as short-term deformations of timber and glulam is very important from a practical point of view. It should have at least the same level of priority as prediction of failure in timbers. Accordingly, a better knowledge of creep in timber is needed, even if the design is rather insensitive to the chosen creep factors in the code.

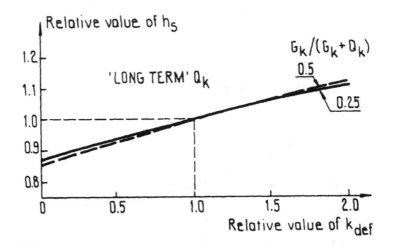

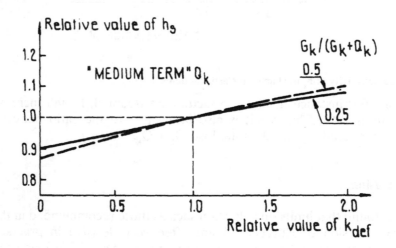

Fig. 1.6 Sensitivity of h_s to change in creep factor k_{def}.

2

Basic knowledge

P. Morlier and L.C. Palka

2.1 Modelling of time-dependent deformations

Engineering design of structural wooden components has relied heavily on experimental results in the past. To predict the magnitude of creep, stress relaxation and rate of damage accumulation, however, numerous mathematical models have been developed for both linear and non-linear materials, including wood.

For linear viscoelastic materials obeying Boltzmann's principle of superposition, series and parallel combinations of elastic springs and viscous dashpots have often been used as a convenient representation. Mathematically linear differential equations relating stress, strain and time can be written so that creep curves, stress–relaxation curves, and ramp-loading curves can be derived one from another (Figs. 2.1, 2.2).

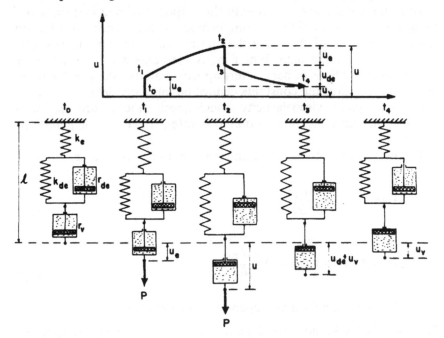

Fig. 2.1 Burger-body representation of creep behaviour. After Bodig and Jayne (1982).

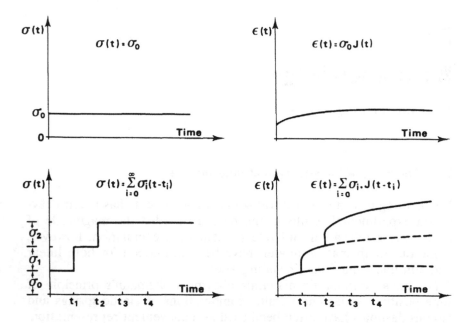

Fig. 2.2 Illustration of Boltzmann's principle of superposition. After Bach and McNatt (1990).

To achieve good agreement with the response of real materials, it is usually necessary to develop complicated models consisting of many simple elements, with fixed retardation time, so that the model response fits the experimental data. When many elements are used, however, various combinations of parameters approximate the data with equivalent precision; thus the constants cannot be uniquely defined.

For non-linear viscoelasticity, mechanical models are no more convenient, and constitutive equations often make use of hereditary integrals.

The hereditary integral for a linear viscoelastic material is

$$\varepsilon(t) = \int_0^t J(t - \tau)\frac{\partial\sigma(\tau)}{\partial\tau}\,d\tau \tag{2.1}$$

or

$$\varepsilon(t) = J_0\sigma(t) + \int_0^t J_d(t - \tau)\frac{\partial\sigma(\tau)}{\partial\tau}\,d\tau \tag{2.2}$$

where J_d contains only time-dependent deformation.

The form of the kernel J (or J_d) may be determined from creep tests. Typical forms of J_d are

Kelvin body $\quad A\left(1 - e^{t/\bar{t}}\right)$

$$(2.3)$$

logarithmic $\quad A \log (t+1)$ $\hspace{4cm}$ (2.4)

$\hspace{2.6cm} A_1 \log (t+1) + A_2 \log^2 (t=1)$ $\hspace{1.5cm}$ (2.5)

$$A \log\left(\frac{1+t}{\tau}\right)$$

$$(2.6)$$

parabolic $\quad A t^m , \; \Sigma \, A_i \, t^{m_i}$ $\hspace{3cm}$ (2.7)

Burger–body $\quad A \, (1 - e^{-t/\bar{t}}) + Bt$

$$\sum_{i=1}^{r} A_i \, (1 - e^{-t/\bar{t}_i}) + B_i t$$

$$(2.8)$$

Non-linear viscoelastic materials may be accurately described by the Green–Rivlin representation

$$\varepsilon(t) = \int_0^t J_1(t - \tau) \frac{\partial \sigma(\tau_1)}{\partial \tau_1} d\tau_1$$

$$= \int_0^t \int_0^t J_2(t - \tau_1, t - \tau_2) \frac{\partial \sigma(\tau_1)}{\partial \tau_1} \frac{\partial \sigma(\tau_2)}{\partial \tau_2} d\tau_1 d\tau_2$$

$$= \int_0^t \int_0^t \int_0^t J_3(t - \tau_1, t - \tau_2, t - \tau_3) \frac{\partial \sigma(\tau_1)}{\partial \tau_1} \frac{\partial \sigma(\tau_2)}{\partial \tau_2} \frac{\partial \sigma(\tau_3)}{\partial \tau_3} d\tau_1 d\tau_2 d\tau_3 \quad (2.9)$$

$$+ \ldots$$

For a creep test, the third order Green–Rivlin formulation reduces to a non-linear polynomial function of stress

$$\varepsilon(t) = \sigma J_1(t) + \sigma^2 J_2(t,t) + \sigma^3 J_3(t,t,t)$$

$$(2.10)$$

It is very long and onerous to launch an experimental programme to calibrate Green–Rivlin formulas, even to the third order only (Fig. 2.3).

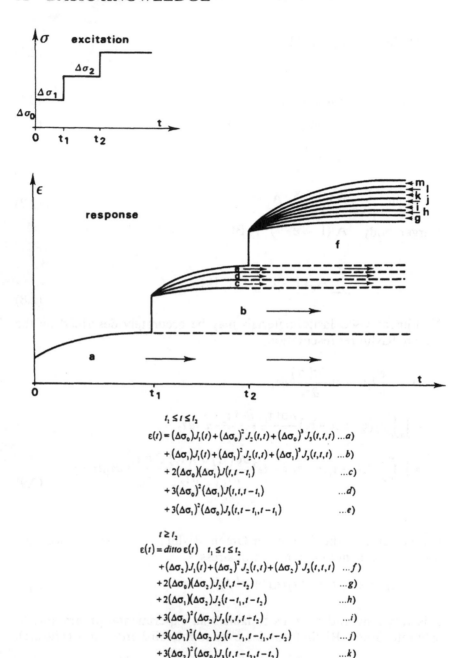

$$t_1 \leq t \leq t_2$$

$$\varepsilon(t) = (\Delta\sigma_0)J_1(t) + (\Delta\sigma_0)^2 J_2(t,t) + (\Delta\sigma_0)^3 J_3(t,t,t) \quad ...a)$$
$$+ (\Delta\sigma_1)J_1(t) + (\Delta\sigma_1)^2 J_2(t,t) + (\Delta\sigma_1)^3 J_3(t,t,t) \quad ...b)$$
$$+ 2(\Delta\sigma_0)(\Delta\sigma_1)J(t,t-t_1) \qquad ...c)$$
$$+ 3(\Delta\sigma_0)^2(\Delta\sigma_1)J(t,t,t-t_1) \qquad ...d)$$
$$+ 3(\Delta\sigma_1)^2(\Delta\sigma_0)J_3(t,t-t_1,t-t_1) \qquad ...e)$$

$$t \geq t_2$$

$$\varepsilon(t) = ditto \; \varepsilon(t) \quad t_1 \leq t \leq t_2$$
$$+ (\Delta\sigma_2)J_1(t) + (\Delta\sigma_2)^2 J_2(t,t) + (\Delta\sigma_2)^3 J_3(t,t,t) \quad ...f)$$
$$+ 2(\Delta\sigma_0)(\Delta\sigma_2)J_2(t,t-t_2) \qquad ...g)$$
$$+ 2(\Delta\sigma_1)(\Delta\sigma_2)J_2(t-t_1,t-t_2) \qquad ...h)$$
$$+ 3(\Delta\sigma_0)^2(\Delta\sigma_2)J_3(t,t,t-t_2) \qquad ...i)$$
$$+ 3(\Delta\sigma_1)^2(\Delta\sigma_2)J_3(t-t_1,t-t_1,t-t_2) \qquad ...j)$$
$$+ 3(\Delta\sigma_2)^2(\Delta\sigma_0)J_3(t,t-t_2,t-t_2) \qquad ...k)$$
$$+ 3(\Delta\sigma_2)^2(\Delta\sigma_1)J_3(t-t_1,t-t_2,t-t_2) \qquad ...l)$$
$$+ 6(\Delta\sigma_0)(\Delta\sigma_1)(\Delta\sigma_2)J_3(t,t-t_1,t-t_2) \qquad ...m)$$

Fig. 2.3 Operation of the Green–Rivlin model. After Whale (1988).

Special assumptions can induce non-linear single integral constitutive equations, for instance:

Nakada (1960)

$$J_2(t,t-t_1) = \sqrt{J_2(t).J_2(t-t_1)}$$

$$J_3(t,t-t_1\ t-t_2) = \sqrt[3]{J_3(t).J_3(t-t_1).J_3(t-t_1)}$$

Gottenberg et al. (1969).

$$J_2(t,t-t_1) = J_2(t+t-t_1) = J_2(2t-t_1)$$

$$J_3(t,t-t_1,t-t_2) = J_3(3t-t_1-t_2)$$

Chueng (1970)

$$J_2(t,t-t_1) = \frac{1}{2}[J_2(t)+J_2(t-t_1)]$$

$$J_3(t,t-t_1,t-t_2) = \frac{1}{3}[J_3(t)+J_3(t-t_1)+J_3(t-t_2)].$$

Alternative non-linear formulations have also been proposed. By assuming that the response ε(t) is non-linear only relative to the present stress value, and not to the previous stress history, Coleman and Noll (1961) gave the following approximation:

$$\varepsilon(t) = J_0\sigma(t) + \int_0^t J[\sigma(\tau),t-\tau]\frac{\partial\sigma(\tau)}{\partial\tau}d\tau.$$
(2.11)

Findley et al. (1967) gave a similar expression, called the modified superposition principle.

According to Schapery (1966), the one-dimensional isothermal creep formulation may be written:

$$\varepsilon(t) = g_0[\sigma(t)]Je\ \sigma(t) + g_1[\sigma(t)]\int_0^t J_1[\psi(t)-\psi(\tau)]\frac{\partial[g_2[\sigma(\tau)]\sigma(\tau)]}{\partial\tau}d\tau$$
(2.12)

where

$$\psi(u) = \int_0^\mu \frac{ds}{h[\sigma(s)]}$$

g_0, g_1, g_2, h are functions of stress.

By assuming $g_1 = h = 1$, Schapery's equation reduces to the modified superposition representation.

A complete discussion of the preceding rheological models can be found in Whale's dissertation (1988).

2.2 Some general observations about creep in wood

As a result of many rheological investigations on wood and wooden materials, much qualitative general information is available.

2.2.1 Limits of linear viscoelasticity in wood

According to Schaffer (1972): 'Wood behaves non-linearly over the whole stress-level range, with linear behaviour being a good approximation at low stresses. Because of this nearly linear response at low levels of stress, Boltzmann's superposition principle applies to stress–strain behaviour for stresses up to 40% of short time behaviour. Much lower stress limits have been observed for wood-based composite materials.'

These trends are supported by an earlier review of creep experiments by Schniewind (1968) and have been confirmed by other studies as well (Bach 1965; Bazant 1985; Nakai and Grossman 1983; Pentoney and Davidson 1962).

In other words 'under long-term loading, when stress, moisture content and temperature are sufficiently low, wood will act essentially in a linear elastic manner; at intermediate values of these variables its behaviour becomes linear viscoelastic in nature, and at higher stress levels, or in fluctuating environmental conditions wood becomes distinctly non-linear viscoelastic in character' (Whale 1988), as illustrated in Fig. 2.4.

2.2.2 Effect of temperature

An increase in temperature will generally reduce the stiffness of timber in bending, tension or compression (Davidson 1962; Bach and McNatt 1990; Kingston and Budgen 1972); especially above 55°C as shown in Fig. 2.5 (Huet et al. 1981; Davidson 1962; Arima 1967). This is known to be the temperature where lignin alters its structure and hemicelluloses begin to soften.

The interaction of creep with variable temperature results in a complex behaviour that may be difficult to predict from constant temperature creep tests (Schniewind 1968).

For example, an increase in temperature in the range 20–90°C during a bending test resulted in a creep that was larger than the creep caused by constant temperature at the highest level (Jouve and Sales 1986); such a trend is evident in Fig. 2.6.

Nevertheless an interpretation of creep curves at different temperatures is given later in this book (Chapter 3) by Arrhenius' law.

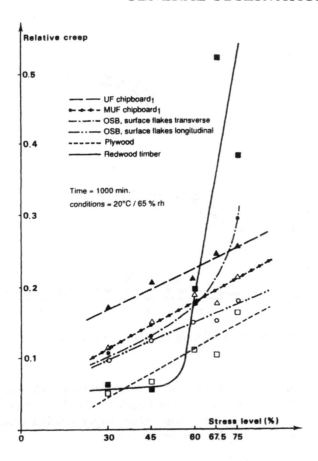

Fig. 2.4 Relation in relative creep with stress level for redwood timber and a number of board types. Points are the means of two to eight samples. After Dinwoodie *et al.* (1990).

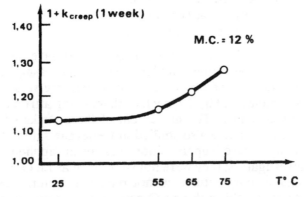

Fig. 2.5 $1+k_{creep}$ (1 week) versus temperature for spruce in bending at constant MC (0.12). In Huet, *et al.* (1981).

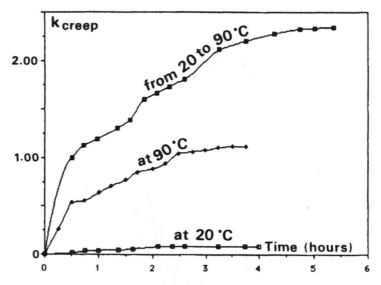

Fig. 2.6 Creep function for a tropical wood. In Huet, *et al.* (1981).

If $J = \bar{\bar{J}}$ (t,T) is the creep function (2 variables) for different values of temperature T, we can find a reduced variable t/τ (T) such as $J = \bar{J}$ (t/τ) (1 variable); this is the **time–temperature superposition principle** and the formulation of τ (T) by Arrhenius' law is

$$\tau(T) = \tau_0 \exp\frac{W}{RT}$$

(2.13)

where   W is an activation energy,
           R is perfect gas constant,
           τ_0 is an empirical parameter.

With this law, Genevaux and Guitard (1988) gave an interpretation of creep tests performed under linearly increasing temperatures. Different peaks can be observed on the creep rate versus temperature curves as depicted in Fig. 2.7, each of them being attributed to a different creep mechanism. The movement of these peaks with changes in the moisture content is also studied in Genevaux's dissertation (1989).

Time–temperature superposition has been applied to wood by several investigators; they were not so positive as Le Govic: 'short-term tests at higher temperatures cannot replace long-term tests at normal temperatures' (Goldsmith and Grossman); 'superposition of time and temperature must be used with caution' (Davidson 1962); however

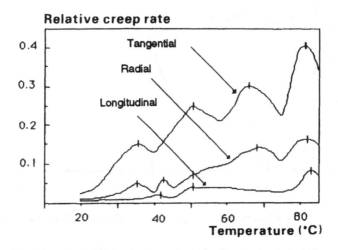

Fig. 2.7 Creep under linearly increasing temperature. Poplar in bending; 3 directions. From Genevaux and Guitard (1988).

Schaffer (1972) concludes that 'the viscoelastic response of wood is thermorheologically simple over narrow temperature ranges'. This last conclusion should not be altered by reading Chapter 3.

2.2.3 Effect of moisture

Moisture in wood acts as a plasticizer. Therefore, creep increases with moisture content (MC) (Bodig and Jayne 1982; Schniewind 1968), as indicated in Fig. 2.8. Bach (1965) deduced from tensile tests, with moisture content ranging from 0.04 to 0.12, that the creep compliance was proportional to the square of moisture content. He also noticed that an increase of 0.04 in moisture content had about the same effect as an increase in temperature of about 6°C within the moisture and temperature limits of his experiment.

Reviews on the effects of the interaction of moisture movement (drying and wetting) with the mechanical behaviour of wood, called the **mechano-sorptive effect** (Grossman 1976, 1978) are provided by Schniewind (1968), Bodig and Jayne (1982), Hunt (1990, 1991) and Gril (1988). Mechano-sorptive creep does not seem to be a particular time-dependent phenomenon, as explained later (Chapter 5).

2.2.4 Anisotropic property of deformation (including creep)

Most creep studies on wood have been conducted in bending, because they are easier to implement and the deformations are much larger and

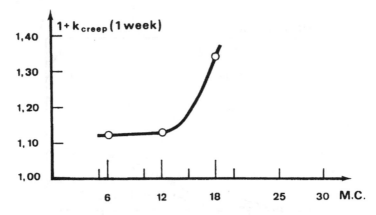

Fig. 2.8 $1+k_{creep}$ (1 week) versus MC for spruce in bending at 25°C. In Huet *et al.* (1981).

easier to measure than for any other strength test. 'Unfortunately, bending tests give structural element properties rather than material properties' (Holzer *et al.* 1984).

Another problem is that most bending tests in the literature are based on three-point loading, such that shearing forces are also present. Since the shear rigidity is usually much less than the bending stiffness in the longitudinal direction, shear creep deformations may not be neglected even for slender beams.

Schniewind and Barrett (1972) report that the relative creep in shear is only 0.6 of that perpendicular to the grain for Douglas-fir specimens loaded for 1000 minutes. On the other hand this relative shear creep is five times larger than relative bending creep parallel to the grain.

Cariou (1987) has determined all nine viscoelastic compliances for a load duration of 1400 minutes, for three moisture contents, on the same species. His results are summarized in Fig. 2.9. The range of his experimental work was doubled by a 'power law' model.

2.2.5 Creep of wood-based materials

Comparative creep studies have shown that relative creep is greatest in fibreboard, followed by particle board, plywood, glued laminated wood and solid wood, relative creep in fibreboard being some five times that of solid wood, under relatively low loads and dry environmental conditions (Fig. 2.4). 'It is interesting to note that the above limited wood-based sheet materials appear in strict descending order as regards the degree to which the whole wood structure is broken down during their manufacture' (Whale 1988) as long as the same low load and dry environmental conditions are maintained.

Exhaustive work by J. Dinwoodie *et al.* over several years (and

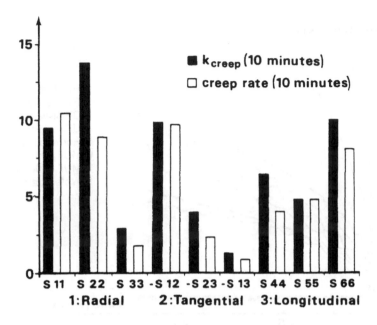

Fig. 2.9 Creep of maritime pine in traction and shear (Cariou 1987).

reported in Dinwoodie *et al.* (1990) for example) demonstrates that the relative order of creep changes for different materials as load levels or environmental conditions (Fig. 2.10) are altered.

Therefore, a general creep ranking of wood-based materials – independent of service loads and environmental conditions – cannot be developed.

2.2.6 Creep-rupture

A discussion of load duration effects in lumber and wood-based structural materials and elements is beyond the scope of this book. For reference, it may be noted that considerable effort has been focused on the examination of creep-rupture, and the associated damage accumulation models, to provide realistic load-duration factors for commercial lumber grades (Foschi *et al.* 1989) and wood-based panel products (Palka 1989a; Laufenberg 1988).

2.3 Some basic questions

A rather fragmented picture of a highly complex creep process has emerged from this literature review. Some basic questions need to be considered here.

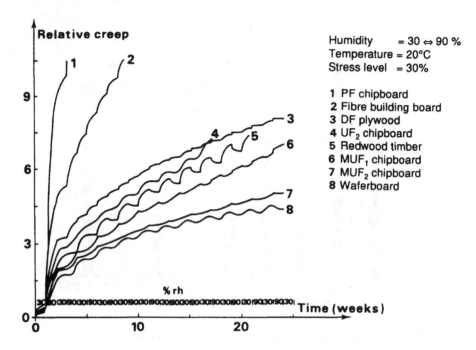

Fig. 2.10 Comparison of relative creep in redwood timber and several board types under cyclic changes in relative humidity. After Dinwoodie *et al.* (1990).

2.3.1 What is the order of magnitude of creep for wood?

The long-term deformation of wooden materials is given by:

$$\varepsilon = \varepsilon_{inst} \left(1 + k_{creep} \right)$$

(2.14)

Where ε_{inst} is measured for a (short) reference time, the term k_{def} is also used, instead of k_{creep}, for example by Thelandersson (Chapter 1) or Ranta-Maunus (Chapter 9).

Depending on the load and environment history, k_{creep} (relative creep) values between 0.5 and 2.0 may be expected for timber at 10 years according to Eurocode 5 (see Chapter 6).

More generally k_{creep} values up to 12 are possible for wood-based panels subjected to unfavourable environmental conditions. Indeed, the long-term creep at failure exceeds the short-term deformation at failure by factors of 1.0 to 2.0 in dry conditions, and by factors of 1.3 to 3.5 in high-humidity chambers, as nominal bending stress levels decrease from 100% to 20% for commercial waferboard panels (Palka and Rovner 1990) .

Table 2.1 k $_{creep}$ values for one year load duration, under naturally varying conditions, collected by Ranta-Maunus (1991)

Meierhofer and Sell (1979)	Glulam, under roof	0.3
Ranta-Maunus (1975)	Glulam, natural environment	0.6
	exposed to rain	1.0
Leivo (1991)	Truss, nail plate connectors, under roof	0.6
Wilkinson (1984)	Truss, under roof	0.9
Nielsen (1972)	Estimation on old buildings	1.3
Ranta-Maunus (1976)	Plywood plates, indoors	1.2
Littleford (1966)	Glulam, indoors	0.3
Bohannan (1974)	Glulam, indoors	0.4
McNatt and Superfesky (1983)	I beam plywood, indoors	0.5
	under roof	0.6

2.3.2 Is it possible to understand the sensitivity of creep relative to changing temperature and moisture content?

The molecular organization of wood is a complex system composed of a large number of chemical substances with molecules of a wide range of sizes and shapes distributed unevenly over the wood fibre walls. Basically, however, three main polymers may be distinguished in wood:

- cellulose, a long-chain semi-crystalline polymer;
- hemicelluloses, medium-length and semi-crystalline polymers;
- lignin, consisting of completely amorphous polymers.

Polymers are known for having a well-defined viscoelastic behaviour. Temperature has a great influence on the deformations either elastic or time-dependent, which are described by the so-called time–temperature superposition principle (extended for given materials to a time–pressure–temperature superposition). The validity of these hypotheses for timber and wood-based composite materials, however, is questionable on the basis of empirical data (Dinwoodie *et al.* 1990). Indeed, moisture content–time superposition was proposed by Douglas and Weitsmann (1980) for composites but it was unsuccessful for wooden materials.

Bodig and Jayne (1982) give a simple description of wood behaviour through an ultra-simplified two-dimensional molecular network (Fig. 2.11.1):

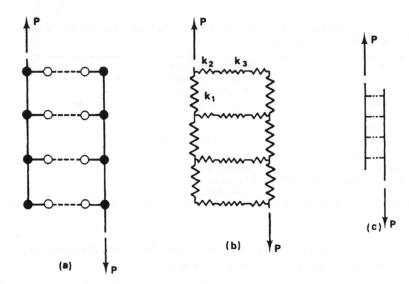

Fig. 2.11.1   A chemical bond model of the rheological behaviour of wood
a) schematic representation
b) equivalent spring representation
c) simplified diagram.

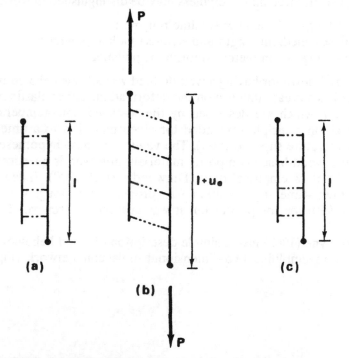

Fig. 2.11.2   Schematic representation of elastic bond deformation:
a) initial condition   b) elastic deformation
c) recovered shape.

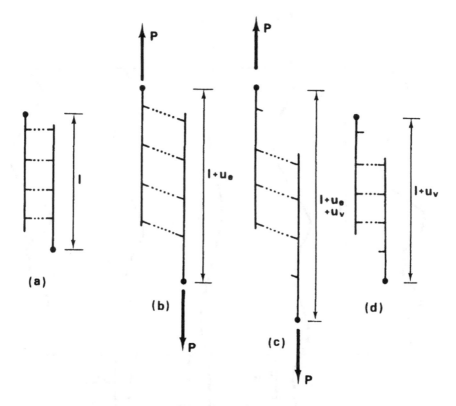

Fig. 2.11.3 Schematic representation of viscous flow:
   a) initial condition   b) elastic deformation
   c) bond breakage   d) newly formed bonds.

- an elastic component (of the deformation) traceable largely to elastic deformation of primary and secondary chemical bonds (Fig. 2.11.2);
- a viscous component (of the deformation) resulting from sections or entire molecules shifting to new positions, through reaching and reforming secondary bonds (Fig. 2.11.3),
- a time-dependent component of the deformations associated with the straightening of molecules and occasioned by secondary bond (hydrogen bonds are predominant here), breakage and reformation (Fig. 2.11.4).

This simplified description makes clear that changing moisture content causes extremely rapid reaction of water with hydrogen bonds (mechano-sorption).

As far as the influence of constant humidity is concerned, it is justifiable to consider, according to Back and Salmen (1982) that adsorbed

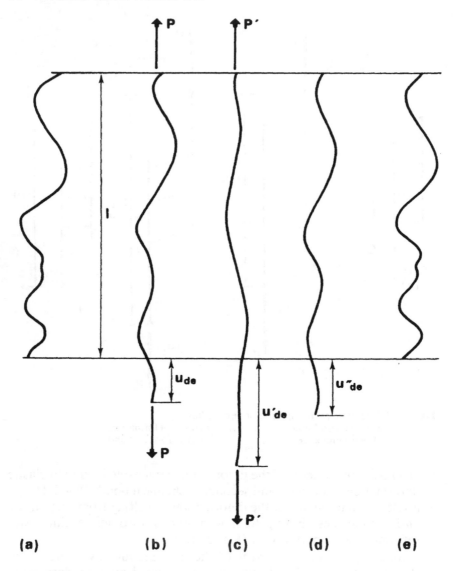

Fig. 2.11.4 Schematic representation of delayed elastic deformation of a polymer chain:
a) original shape
b) early stage of deformation
c) final stage of elongation
d) early stage of recovery
e) recovered shape.

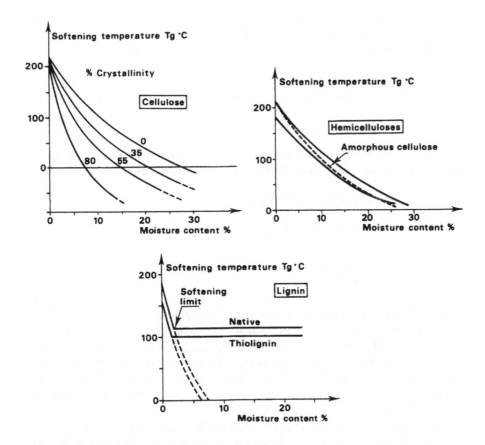

Fig. 2.12 Plastification of isolated wood components after Back and Salmen (1982).

water in the polymers produce a variation of the transition temperature of these polymers as indicated in Fig. 2.12. The result is an activation effect linked to increasing moisture content.*2.2.3 What do we know about mechano-sorptive creep?*

Hunt gives (Chapter 5) an exhaustive state-of-the-art review of mechano-sorptive effects in wood and questions the existence of a theoretical creep limit.

2.3.4 Does a mechano-thermic creep exist, with the same consequence?

Researchers who carefully analyse time–temperature superposition affirm that this is not the case (Genevaux 1989).

2.3.5 Finally is it possible to predict creep of wood?

In the following section an effort is made to give general information about the classical modelling of creep, and to examine the possibility of describing correctly an observed deformation, or predicting it.

2.4 Predictive models for creep in wood and wood-based materials

2.4.1 Under constant climate

The basic paper was written by Gressel (1984); having experimental data for creep duration up to ten years, he tried four different creep functions:

$$(a) \quad \varepsilon(t) = \beta_1 + \beta_2\left(1 - e^{-\beta_3 t}\right) + \beta_4 t$$

$$(b) \qquad = \beta_1 + \beta_2\left(1 - e^{-\beta_3 t}\right)$$

$$(c) \qquad = \beta_1 t^{\beta_2} + \beta_3$$

$$(d) \qquad = \beta_1 t^{\beta_2} \tag{2.15}$$

'The adaptation precision of ten years' experiments proved excellent using the chosen potential functions for the 4-parameter model, but was clearly less relevant for the 3-parameter model. Extrapolation beyond the test duration supplied excessively high estimated deformations with the 4-parameter model, and too low values with the 3-parameter model. When both potential functions yielded about the same values, plausible prediction was reached', according to Gressel (1984).

For commercial waferboard panels in bending, 4-parameter creep models (function (a) above) have been fitted to about 600 specimens representing 20 different combinations of constant loads and environmental conditions (Palka and Rovner 1990). Analyses revealed these parameters to be highly correlated with both load levels and environmental conditions, as well as with each other. Evidently, these four waferboard creep parameters are not the independent material characteristics implied by function (a). Indeed, each parameter in a simplified creep model represents a different and complex (unknown) interaction of many basic physical characteristics that affect the observed elastic, viscoelastic and viscoplastic behaviour of composite wood-based materials.

Over the last 15 years, extensive work on quantifying and predicting creep behaviour of particle board has been undertaken by Dinwoodie *et al.* (1990). They concluded that 'although the 3-element (function (b)) model is capable of accurately describing creep for short periods of time, it is of little use in the prediction of long-term creep from short-term data, since it, incorrectly, assumes no further creep beyond the last data point ... The 4-element model (function (a)) predicted too high a defection; since it assumes a constant rate of viscous creep after the last data point . . .' For Dinwoodie *et al.* (1984), it became necessary to resort to a 5-parameter model, with a non-linear viscoelastic dashpot, to have good correlation with experimental data:

(e) $\varepsilon(t) = \beta_1 + \beta_2\left(1 - e^{-\beta_3 t}\right) + \beta_4 t^{\beta_5}$

$$(2.16)$$

Mukudai (1983) used an 8-parameter model (one Maxwell element followed by three Kelvin elements in series) to describe bending creep in Japanese cypress.

For describing and predicting creep, linear models quickly become unsatisfactory. Van Der Put (1989, Chapter 4) applies deformation kinetics, a previous theory from Krausz and Eyring, to wood. He analyses the response of a non-linear 3-element model, including a non linear dashpot

$$\dot{\varepsilon} = A \sinh(\varphi\sigma) \tag{2.17}$$

to different loading (constant strain rate test, constant loading rate test, creep, stress relaxation) and he particularly demonstrates that creep curve (e – log t) has a delayed acceleration, from a small slope to a steeper one; he also explains that it is possible to derive and discuss the LWF -equation, giving the shift factor of the displacement of the curve along the log–time axes in the time–temperature superposition principle, as demonstrated in Fig. 2.13.

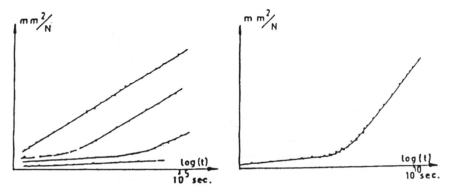

Fig. 2.13 Van Der Put's model and creep results for fibre bundles. Master creep curve is obtained by moving the curve along the horizontal log(t) axis.

In Van Der Put's work, a master curve is also illustrated for creep tests, underlining a stress–time equivalence (or superposition). This type of equivalence, like the time–temperature superposition often referred to by Le Govic and Huet *et al.* (1981), may be the best and physically consistent way to predict long term creep results from short–term experiments, as shown in Fig. 2.14; provided the model parameters used in estimating this time–temperature equivalence are truly (reasonably) independent of each other and of changing temperatures or load levels and duration.

In his dissertation, Whale (1988) performs a very interesting exercise concerning non-linear models: he makes use of numerous embedment tests and chooses twelve rheological models. These models are first calibrated through a multi-step creep and recovery loading regime, then a validation is tried on intermittent or varying loading regimes (Fig. 2.15.1).

The scattering of the different model responses was unexpected (Fig. 2.15.2). The model chosen by Whale (new hybrid joints), however, seems to give good fit to experimental data. His results are reviewed again in the chapter devoted to time-dependent slip of joints.

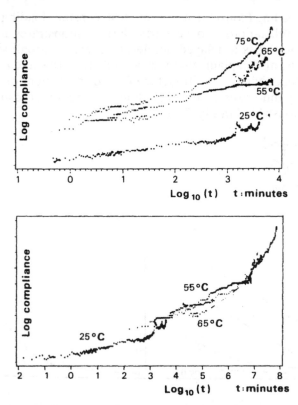

Fig. 2.14 Creep compliance at different temperatures and master curve (spruce in bending – CTBA data).

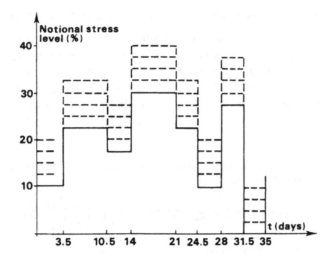

Fig. 2.15.1 Validatory loading regime.

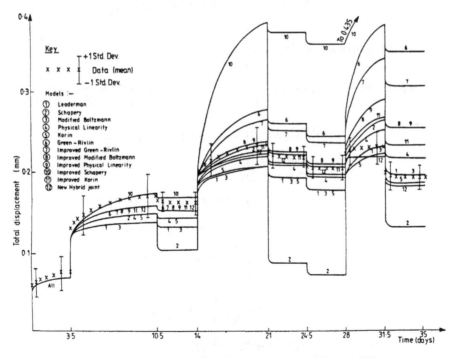

Fig. 2.15.2 Mean experimental and predicted responses. After Whale (1988).

2.4.2 Under constant but different environmental conditions

Different authors, for instance Bach (1965), Pierce and Dinwoodie (1977), Cariou (1987) have given relationships between the parameters of creep function and specimen temperature and moisture level.

Table 2.2 given by Palka summarizes possible simplified models for describing the effects of constant but different environmental conditions on creep under constant external load. 'For the power functions,

Table 2.2 Effects of environmental conditions on creep under constant load (ε_o).
Power function (4 parameters):

$$\varepsilon(t,\sigma_o,M,T) - \varepsilon_{EN}(M,T) = \left\{ A_o \left[1 + \left(\frac{t}{t_r} \right)^r \right] \right\} \frac{\sigma_o}{a_{MT}}$$

Schapery function (7 parameters):

$$\varepsilon(t,\sigma_o,M,T) - \varepsilon_{EN}(M,T) = \left\{ g_o D_o + g_1 \left[g_2 D_1 \left(\frac{t}{a_\sigma} \right)^n \right] \right\} \sigma_o$$

Spring-and-dashpot analog (7 parameters):

$$\varepsilon(t,\sigma_o,M,T) - \varepsilon_{EN}(M,T) = \left\{ J_o + J_1 \left(1 - e^{\frac{-t}{Q_1}} \right) + J_2 t \left[1 + 2 \left(\frac{t}{Q_2} \right)^s \right] \right\} \frac{\sigma_o}{a_{LI}}$$

Where:

$$\varepsilon_{EN}(T,M) = \varepsilon_o \left\{ [1 + \alpha_M . \Delta M] \mathrm{x} [1 + \alpha_T . \Delta T] \right\}$$

$$\Delta M = M_o - M$$

$$\Delta T = T_o - T$$

$$g_i = g_{oi}(\sigma_o, M_o, T_o) \mathrm{x} f_i(\sigma_o, M, T)$$

$$a_\sigma = a_{\sigma o}(\sigma_o, M_o, T_o) \mathrm{x} g_i(\sigma_o, M, T)$$

$$a_{MT} = \exp \left\{ r \left[\frac{W_M}{R_M . \Delta M} + \frac{W_T}{R_T . \Delta T} \right] \right\}$$

$$= \exp \left\{ \frac{r}{R} \left[\frac{W_M}{\Delta M} + \frac{W_T}{\Delta T} \right] \right\}$$

$$a_{LI} = a_\sigma . a_E \approx a_{MT}$$

Here, β and γ_i are linear moisture–temperature effect parameters (know for small clear wood specimens), while r, R, R_i, W_i are non-linear moisture temperature effect parameters (yet unknown); α_i are moisture and heat expansion coefficients; g_i and α_o are unknown moisture and temperature-dependent (empirical) parameters of the Schapery function.

the combined or independent effects of changing equilibrium moisture content and temperature may be estimated from modified classical rate equations' (the same role is given to temperature and moisture content, which is controversial). 'The Schapery function assumes that the horizontal and vertical shift factors . . . are basic material parameters. The modified Burger-body model incorporates the simplifying assumption that all compliances exhibit the same moisture–temperature effects'.

2.4.3 Under varying environmental conditions

We come back to the difficult problem of mechano-sorptive creep for which two different approaches are studied (Toratti 1991a).

The linear model, presented by Ranta-Maunus (1975) is as follows:

$$\varepsilon_{ms} = \int [\alpha + m\sigma]\, du(\tau) \tag{2.18}$$

where α is the shrinkage-swelling coefficient, m is a mechano-sorptive material parameter having different values for sorption, m+, or m++ (sorption to moisture contents not previously attained during loading history) or desorption m⁻.

'Good agreement has been found between the model and tests results, since mechano-sorption has been found to be proportional to the moisture content change. Although, if the load duration is extrapolated over the period of the test, from the period from which the material parameters were derived from, there is reason to doubt if the material parameters are valid any more, or if there remains linearity between strain and moisture content change' (Toratti 1991a).

For the non-linear model, proposed by Hunt and Shelton (1988) the equation is given in two parts:

one for reversible strain $\varepsilon_{rev} = \int [\propto -b\varepsilon(t)] du(\tau)$

and the other for permanent strain $\varepsilon_{ms} = J^{\infty}\sigma\left[1 - \exp\left(-c\int |du(\tau)|\right)\right]$ $\tag{2.19}$

The reversible moisture-induced strain is affected by the strain state (parameter b), which accounts for the oscillation of the creep curve when the specimen is subjected to moisture content cycling. The mechano-sorption strain is assumed to have a limit value determined by a moisture accumulation parameter c.

A. Martensson (1990) has presented a sophisticated model which was quantified on the basis of creep tests in cyclic climate and then checked against other types of tests.

There is, however, a lot of further work to do in order to get a complete description and a good prediction of mechano-sorptive deflections in timber constructions.

Table 2.3 Creep tests in constant environments: wood specimens

Specimen identification	Load type	Stress level	Load duration	Moisture	Temperature	Creep data	References
Domestic and pical species	Tensile ||	6 to 86 % *	30 minutes	12 % MC	75 to 80°F	Equation (2.4)	King (1961)
Hard maple (3.5 x 4.5 x 47 mm)	Tensile ||	20 to 80 % *	1000 minutes	4 to 12 % MC	30 to 70°C	Equation (2.5)	Bach and Pentoney (1968)
Douglas-fir (1 x 1 x 22 in.)	Bending ||	60 to 85 % *	10 years	6 and 12 % MC	75 & 80°F	Equation (2.7)	Clauser (1959)
Douglas-fir	Twisting and tensile	linear	1000 minutes	10 % MC	72°F	Equation (2.7)	Schniewind and Barrett (1972)
Douglas-fir (89 x 89 x 4900 mm)	Bending ||	linear	400 hours	12 % MC	70°F	Equation (2.7)	Hoyle et al. (1986)
Douglas-fir (2 x 4 x 96 in.)	Bending (4 points)	98, 88, and 67 % *	220 days	55% RH (9 % MC)	75°F	Equation (2.7) Measured deflection	Gerhards (1985)
Douglas-fir (Tension: .25 x .75 x 7.5in) (Shear: 30 x 9 x 9 in.)	Tension ⊥ Plate shear	15 to 25 % *	1000 minutes	50% RH (10 % MC)	72°F	Equation (2.7) Measured deflection and angle	Schniewind and Barrett (1972)
Sitka spruce	Bending || (3 points)	15 to 50 % *	8 hours	11 % MC	40°C	Equation (2.7) Measured deflection and angle	Urakami and Fukuyama (1982)
Spruce, Pine, Beech	Bending ||	linear	10 years	12 % MC green & air dry	20°C	Equation (2.7)	Gressel (1984)
Hoop pine (0.8 x 0.8 x 36 in.)	Bending ||	linear and non-linear	50 days		21°C	Equation (2.8) with r = 3, 4	Grossman and Kingston (1954)
White pine, Red cedar and Sugar maple (10 x 20 x 300 mm)	Bending ||	linear	6 to 10 hours	15.8 to 20.5 % MC	20 to 60°C	Equation (2.8) with r = 40 without flow term	Davidson (1962)
Teak (3.2 x 9.5 x 64 mm)	Tensile ||	linear and non-linear	8 days	12.5 % MC	30°C	Equation (2.8) with r = 2 without flow term	Bhatnagar (1964)
White spruce	Compression ||	linear and non-linear		15 % MC		Equation (2.8) with r = 1 without flow term	Ylinen (1965)
Sitka spruce (2 x 2 x 8 in.)	Compression ||	10 to 60 % *	20 days	93 % MC	31°C	Equation (2.8) with r = 1	Senft and Suddarth (1971)

Table 2.3 cont.

Specimen identification	Load type	Stress level	Load duration	Moisture	Temperature	Creep data	References
Sitka spruce (small clear dimension)	Bending || (3 points)	linear	7 hours	60% RH	20°C	Equation (2.8) with r = 1	Urakami and Fukuyama (1982)
Hinoki (5 x 15 x 150 mm)	Bending || (4 points)	30 to 60 % *	10 hours	11 % MC	25°C	Equation (2.8) with r = 3	Mukudai (1963)
Spruce, Pine, Beech	Bending |	linear	10 years	12 % MC	20°C	Equation (2.8) with r = 1	Gressel (1984)
Southern pine n° 2 (2 x 10 x 192 in.)	Bending ||	linear	110 days	55% RH (9 % MC)	75°F	Equation (2.8) with r = 1	Leichti and Tang (1989)
South African pine (38 x 114 x 4000 mm)	Bending	linear	300 days	Ambient laboratory conditions		Equation (2.8) with r = 1 (modified)	Pienaar and Bronkhorst (1982)
Cunningham pine Mountain ash Jarra	Bending || (3 loading cycles)	17, 33, 50 and 67 % *	190 days	60% RH	25°C	Measured deflection	Nakai and Grossman (1983)
Spruce (10 x 25 x 500 mm)	Bending || (4 points)	10, 20, 30 and 40 %	1 week	12 % MC	25, 55, 65,75°C	Equation (2.7) plus thermoactivation law	Le Govic et al. (1988)
Tropical hardwood	Bending || (isostress)	25, 30 and 35 %	2 weeks	12, 15, 18 % MC	25°C	Equation (2.8)	Foudjet (1986)
Maritime pine	All the visco-elastic compliances (tension, shear tests)	linear	1 day	8, 10, 20 % MC	25°C	Equation (2.7)	Cariou (1987)
Spruce (20 x 50 x 1000 mm)	Bending || (4 points)	10, 20, 25, 30, 35 and 40%	2 weeks to 47 weeks	18 % MC	25°C	Equation (2.7)	Rouger et al. (1990)

* of ultimate short-term strengths

Table 2.4 Creep tests in changing environments: wood specimens

Specimen identification	Load type	Stress level	Load duration	Moisture	Temperature	Creep data	References
Spruce wood (small clear dimensions)	Compression ∥ and ⊥ Tension ∥ and ⊥ Shear ∥	70%* and at zero creep	1 to 2 hours and 1 hour	45% RH and 85% RH	-25, 5, 25, 50, 70°C	Equation (2.6)	Van Der Put (1989)
Spruce wood (25 x 50 x 50 mm)	Tension ⊥	Linear	3 days to 3 years	65% RH	20°C (30 and 40°C)	Measured deflection	Kolb et al. (1985)
Differents species and sizes (review)	Differents tests (Results summary)	Varied	Varied	Varied	Varied	Summary of general trends (references)	Schniewind (1968)
Differents species and sizes (review)	Bending, tension and shear	Linear and non-linear	Varied	0 to 30% MC	20°C ±	'Hydroviscoelastic' equations	Ranta-Maunus (1975)
Beechwood (small clear dimensions)	Tension ⊥	Linear	1000 hours	0 to 30% MC	20, 30, 40, 50 and 60°C	Equation (2.7)	Schniewind (1966)
Beechwood (radial) (3 x 6 x 130 mm)	Tension ⊥	20%, 30%, 40% and 50%*	–	Cyclic (1.25-1% MC)	20°C	Empirical measured deflection	Molinski (1987)
Douglas-fir, N° 2+ (3.5 x 3.5 x 96 in.)	Bending ∥ (5 points)	Linear (2 levels)	1200 hours	Cyclic (7 to 20% MC)	70°F	Equation (2.7) and (2.8) with r = 1	Hoyle et al. (1986)
White spruce (20 x 20 x 500 mm)	Bending ∥ (3 points)	7% to 16% *	150 days	Protected external conditions		Equation (2.8) with r = 1 and no flow	Pozgai (1986)
Douglas-fir (0.4 x 0.8 x 8 in.) (2 x 2 x 40 in.)	Bending ∥ (3 points)	70% *	1000 hours	Cyclic (35% to 87% RH)	20°C	Measured deflection	Schniewind and Lyon (1973)
Douglas-fir and hemlock (2 x 6 x 144 in.)	Bending	Linear (2 levels)	3 months to 3 years	8% to 14% MC	10 to 30°C	Measured deflection	–
Scotch pine (2.5 x 2.5 x 7 mm)	Compression ∥	Linear (4 levels)	720 hours	Five cycles (3% to 27% MC)	20°C	Equation (2.7)	Molinski et al. (1987)
Scotch pine (10 x 25 x 450 mm)	Bending ∥ (4 points)	15%, 30%, 45% and 60% *	–	8% MC initial plus moistening	–	Measured deflection	Molinski and Raczkowski (1988)
Beech and mahogany (2 x 2 x 60 mm)	Bending ∥ (3 points)	25% *	55 days	30% RH and 90% RH	20, 25, 30 and 40°C	Measured deflection	Hearmon and Paton (1964)
Yellow poplar (3/8 x 1/2 x 32 in.)	Bending ∥ (4 points)	30% and 15% 56% and 28% *	10 days	6% and 24% MC	80°F	Equation (2.8) with r = 1 Measured deflection	Szabo and Ifju (1970)
(mainly) poplar	Bending L, R, T (isostress)	Linear	400 hours	20% MC	$\dot{T} = 0.2°C\ min^{-1}$	Equation (2.8)	Genevaux (1989)

* of ultimate short-term strengths

Table 2.5 Creep test in changing environments: panel products

Specimen identification	Load type	Stress level	Load duration	Moisture Temperature	Creep data	References
Laboratory waferboard (13 x 125 x 550 mm)	Bending ‖(3 points)	Linear (1.5 x design)	84 days	Different environmental conditions (three)	Equations (2.7) and (2.8) with r = 1 Measured deflection	Alexopoulos (1989)
Commercial waferboard (16 x 300 x 1220 mm) Creep and Creep rupture	Bending ‖(3 points)	15, 30, 55, 65 and 75% *	Current : 2 years Planned : 3 years	Different environmental conditions (five)	Equations (2.8) with r = 1 Measured deflection	Palka (1989a) Palka and Rovner (1989)
Oriented strand board (6.5 x 300 x 1220 mm)	Bending ‖ and ⊥ Compression ‖ and ⊥	Linear	90 days	Controlled chamber (70% RH, 20°C) and uncontrolled interior environment	Measured deflection	Wong et al. (1988)
Oriented strand board (16 x 300 x 1000 mm) Creep and Creep rupture	Bending ‖(3 points)	15, 30, 55, 65, 75 and 80% * †	6 months	Constant dry and cyclic environmental conditions	Measured deflection	Laufenberg (1985) Simulski (1989)
Soft wood plywood (16 x 300 x 1000 mm) Creep and Creep rupture	Bending ‖(3 points)	15, 30, 75 80 and 85% * †	6 months	Constant dry and cyclic environmental conditions	Measured deflection	Laufenberg (1985) Simulski (1989)
Soft wood plywood (15 x 125 x 550 mm)	Bending ‖(3 points)	Linear (1.5 x design)	84 days	Different environmental conditions	Equations (2.7) and (2.8) with r = 1 Measured deflection	Alexopoulos (1989)
Structural panels Literature review	Bending ‖ (3 and 4 points)	Linear to non-linear	Varied	Different environmental conditions	Measured deflection	Laufenberg (1985)
Structural panels (13–18 x 300 x 300 mm) Plywood, Chipboards, Waferboards	Bending ‖(4 points)	60% *	6 months	Different environmental conditions (five, constant)	Equation (2.8) with r = 1 Measured deflection	Dinwoodie et al. (1990)
Structural panels Plywood, Chipboards, Waterboards	Bending ‖ (Varied)	Linear	Varied	Different environmental conditions	Equation (2.7) or alternate forms	Arima et al. (1986)

* of ultimate short-term strengths
† Tests performed but not analysed yet

Table 2.6 Creep tests in changing environments: structural components

Specimen identification	Load type	Stress level	Load duration	Moisture	Temperature	Creep data	References
Wood columns: Douglas-fir N° 2 (4 x 4 x 192 in.)	Compression ∥ (with eccentric loads)	70% *	450 hours	12% MC	60 to 95 °F	Equation (2.7) and (2.8) with r = 1, no flow	Hani et al. (1986)
Prestressed beams: Douglas-fir $\left(3\frac{1}{4} \times 5\frac{1}{16} \times 102 \text{ in.}\right)$	Bending ∥ (4 points)	Linear	3000 days (8 years)	Ambient laboratory conditions		Measured deflection prestresses	Bohannan (1974)
Glulam beam: Douglas-fir (3.5 x 10 x 192 in.)	Bending ∥ (centre point)	Linear (2-2.25 x design)	700 days	Ambient (heated) laboratory conditions		Measured centre deflection	Littleford (1966)
Glulam beam: Taiwan cypress (4 x 6 x 150 cm)	Bending ∥ (cantilevered)	84, 70, 56, 42 and 28% *	1000 days	Ambient (heated) laboratory conditions		Measured end deflection	Sasaki and Maku (1963)
Glulam girders: (97 x 129 x 3080 mm)	Bending ∥ (4 points)	Dead and snow Dead alone	75 days 285 days	Protected exterior environment		Measured deflection strain and MC	Badstube et al. (1989)
Glulam beams: Alpose spruce (139 x 290 x 4000 mm)	Bending ∥ (3 points)	30 and 60% *	8 to 28 months	Ambient laboratory conditions		Measured deflection	Zaupa (1989)
Composite I beams: Southern pine / waterboard (2.6 x 10 x 192 in.)	Bending ∥ (3 points)	15 and 25% *	110 days	55% RH	75°F	Equation (2.8) with r = 1	Leichti and Tang (1989)
Composite I beams: LVL lumber / hardboard LVL lumber / plywood (1 x 12 or 1 x 6 ft)	Bending ∥ (loaded at 2 ft intervals)	Linear	2 years and 5 years	Uncontrolled interior environment Protected exterior environment Cyclic humidity conditions		Measured deflection	McNatt and Superfesky (1983)

Table 2.6 cont.

Specimen identification	Load type	Stress level	Load duration	Moisture	Temperature	Creep data	References
Stressed skin panels : SPF lumber / OSB boards (77 x 455 x 2240 mm)	Bending || (4 points)	Linear	90 days	Controlled chambers (70% RH, 20°C) and uncontrolled interior environment		Measured deflection	Wong *et al.* (1988)
Timbers fasteners : Literature survey of timbers connections	Lateral resistance (tension and compression)	Linear and non-linear	up to 4000 days	Different environmental conditions		Equations (2.7) and (2.8) with r = 1, no flow and others. Measured slip	Palka (1981)
Nailed or bolted joints : Model review and calibration for European redwood	Lateral resistance (tension and compression)	5% to 45% *	up to 35 days	Ambient laboratory conditions		Measured deflection or embedment	Whale (1988)
Timber joints : Four joints types	Tension ||	Linear (1.0 x design)	800 days	Controlled cyclic and constant humidity conditions		Measured joint slip	Feldborg (1989)
Truss-plate joints : Creep and Creep rupture	Lateral resistance in tension	60 and 85% *ù	3 years	Ambient laboratory conditions		Equation (2.8) with r = 1 Measured joint slip	Palka (1989b)
Trusses : Review of Creep and Creep data	Bending	Linear and non-linear	up to 10 years	Different environmental conditions		Measured deflections	Palka (1981)
Trusses : Review of Creep	Bending	Linear and non-linear 10, 30%	up to 10 years	Different environmental conditions		Measured deflections	Arima *et al.* (1986)
French softwood and poplar (40 x 100 x 2000 mm) Glulam (100 x 400 x 7500 mm)	Bending || (4 points)		14 months	12%, 18% MC	25°C	Equation (2.7)	Rouger *et al.* (1990)

* of ultimate short-term strengths

2.5 References

Alexopoulos, J. (1989). Effect of resin content on creep and other properties of wafer-board. Master of Science in Forestry Thesis. University of Toronto, Ontario, Canada, p.126.

Arima, T. (1967). The influence of high temperature on compressive creep of wood. *Jap. Wood. Res. Soc.* **13**(2) 36–40.

Arima, T., Marnyama, N. and Sato, M. (1986). Creep of wood, wood based material and wood composite elements at various conditions. Technical Paper, The Eighteenth IUFRO World Congress, Ljubljana, Yugoslavia, p. 24.

Bach, L. (1965). Non-linear mechanical behaviour of wood in longitudinal tension. Ph. D. Dissertation. Syracuse University, New York, p. 250.

Bach, L. and Pentoney, R.E. (1968). Nonlinear mechanical behavior of wood. *For. Prod. J.* **18**(3) : 60–66.

Bach, L.and McNatt, J.D. (1990). Creep of OSB with various strand alignments. IUFRO S5.02 Meeting St John.

Back, E.L. and Salmen, L. (1982). Glass transitions of wood components hold implications for molding and pulping processes. *TAPPI* 65(7).

Badstube, M., Rug, W. and Schone, W. (1989). Long-term tests with glued laminated timber girders. *International Council for Building Research and Documentation* (CIB W18), Meeting Twenty-Two. East Berlin, GDR, p. 23.

Bazant, Z.P. (1985). Constitutive equation of wood at variable humidity and temperature. *Wood Sci. Technol.* **19** 159–177.

Bhatnagar, N.S. (1964). Creep of wood in tension parallel to grain. *Holz als Roh-und Werkstoff.* **22**(8) 296–299

Bodig, J. and Jayne, B.A. (1982). *Mechanics of wood and wood composites.* Van Nostrand Reinhold, New York, p. 712.

Bohannan, B. (1974). *Time-dependent characteristics of prestressed wood beams.* Research Paper FPL 226. USDA, Forest Service, Forest Products Laboratory, Madison, WI, p. 9.

Cariou, J.L. (1987). Caractérisation d'un matériau viscoélastique anisotrope, le bois. Doctoral Thesis, University of Bordeaux, France.

Chueng, J.B. (1970). Nonlinear viscoelastic stress analysis of blood vessels. Ph. D. Thesis, University of Minnesota, USA.

Clauser, W.S. (1959). *Creep of small wood beams under constant bending load.* Report 2150. USDA, Forest Service, Forest Products Laboratory, Madison, WI, p.18.

Coleman, B.D.and Noll, W. (1961). Foundations of linear viscoelasticity. *Reviews of Modern Physics,* **33**.

Davidson, R.W. (1962). The influence of temperature on creep in wood. *For. Prod. J.* **12**(8) 377–381.

Dinwoodie, J.M., Pierce, C.B. and Paxton, B.H. (1984). Creep in chipboard (Part IV). *Wood Sc. Technol.,* **18**.

Dinwoodie J.M., Higgins, J.A., Paxton, B.H. and Robson, D.J. (1990). Creep research on particle board. *Holz als Roh-und Werkstoff,* **48**.

Douglas, D.A. and Weitsmann,Y. (1980). Stresses due to environmental conditioning of cross-ply graphite/epoxy laminates. *ICCM* 3, 1.

Feldborg, T. (1989). Timber joints in tension and nails in withdrawal under long-term loading and alternating humidity. *For. Prod. J.* **39**(11/12) 8–12.

Findley, W.N. and Lai, J.S.Y. (1967). A modified superposition principle applied to creep of non-linear viscoelastic material under abrupt changes in state of continued stress. *Trans. Soc. Rheology,* 11-3.

Foschi, R.O., Folz, B.R. and Yao, F.Z. (1989). *Reliability-based design of wood structures.* Structural Research Series, Report 34. Dept. of Civil Engineering, University of British Columbia, Vancouver, Canada, 282p.

Foudjet, A. (1986). Contribution à l'étude rhéologique du matériau bois, Doctoral Thesis, University of Lyon, France.

Genevaux, J.M. (1989). Le fluage à température linéairement croissante: caractérisation des sources de viscoélasticité anisotrope du bois, Doctoral Thesis, Institut National Polytechnique, University of Nancy, France.

Genevaux, J.M., and Guitard, D. (1988). Anisotropie du comportement différé; essai de fluage à temperature croissante d'un bois de peuplier. *Colloque Mechanical Behaviour of Wood*, Bordeaux, France.

Gottenberg, W.G., Bird, J.O. and Agrawal, G.L. (1969). An experimental study of a non-linear viscoelastic solid in uniaxial tension. *J. Appl. Mech.*, **36**-E.

Gressel, P. (1984). Prediction of long-term deformation behaviour from short-term creep experiments. *Holz als Roh-und Werkstoff.* **42**(8) 293–301.

Gril, J. (1988). Une modélisation du comportement hygro-rhéologique du bois à partir de sa microstructure. Thèse, Université PARIS VI, France.

Grossman, P.U.A. (1976). Requirements for a model that exhibits mechano-sorptive behaviour. *Wood Sc. Tech.*, **10**, 163–168.

Grossman, P.U.A. (1978). Mechano sorptive behaviour in general constitutive relations for wood and wood based materials. Workshop WSF, Syracuse University, USA.

Grossman, P.U.A. and Kingston, R.G.T. (1954). Creep and stress relaxation in wood during bending. *Aust. J. Appl. Sci.* **5**(4) 403–417.

Hani, R.Y., Griffith, M.C. and Hoyle, R.J. Jr. (1986). The effect of creep on long wood column design and performance. *J. Structural Engineering*, **112**(5) 1097–1114.

Hearmon, R.F.S. and Paton, J.M. (1964). Moisture content changes and creep of wood. *For. Prod. J.* **4**(8) 357–359.

Holzer, S.M., Loferski, J.R. and Dillard, D.A. (1984). A review of creep in wood: concept relevant to develop long-term behaviour predictions for wood structures. *Wood and Fiber Science*, **21**(4) 376–392.

Hoyle, R.J. Jr., Hani, R.Y. and Eckard, J.J. (1986). Creep of Douglas-fir beams due to cyclic humidity fluctuations. *Wood and Fiber Science.* **18**(3) 468–477.

Huet, C., Guitard, D. and Morlier, P. (1981). Le bois en structure, son comportement différé. *Annales I.T.B.T.P.*, **470**.

Hunt, D.G. and Shelton, C.F. (1988). Longitudinal moisture – shrinkage coefficients of softwood at the mechano-sorptive creep limit. *Wood Sci. Technol.* **22** 199–210.

Jouve, J.H. and Sales, C. (1986). Influence de traitements physico-chimiques sur le fluage du matériau bois. Séminaire G.S. Rhéologie du Bois.

King, E.J.,Jr. (1961). Time-dependent strain behaviour of wood in tension parallel to grain. *For. Prod. J.* **11**(3) 156–165.

Kingston, R.S.T. and Budgen, B. (1972). Some aspects of the rheological behaviour of wood. *Wood. Sc. Tech.*, **6**.

Kolb, H., Goth, H. and Epple, A. (1985). Influence of long-term loading, temperature and climate changes on the tensile strength perpendicular to grain of spruce. *Holz als Roh-und Werkstoff.* **43**(1985) 4463–4468.

Korin, U. (1986). *Non-linear creep superposition.* International Council for Building Research and Documentation, Working Commission W18 – Timber Structures. 8p.

Laufenberg, T.L. (1985). Creep and creep rupture in reconstituted panel products. A cooperative study by Forintek Canada Corp. and Forest Products Laboratory, USDA. In: *Proceedings of International Workshop on Duration of Load in Lumber and Wood Products*, Richmond, B.C. Forintek Canada Corp. Vancouver, B.C. p. 61–66.

Laufenberg, T.L. (1988). Composite product rupture under long-term loads: a technology assessment. *Proceedings, XXII International Particle board/Composite Materials Symposium*, Washington State University, Pullman, W.A., p. 241–256.

Le Govic, C., Hadj Hamou, A., Rouger, F.L. and Felix, B. (1988). Modélisation du fluage du bois sur la base d'une équivalence temps-température. *Actes du 2ème Colloque Sciences et Industries du Bois*, Nancy (avril 1987), A.R.B.O.L.O.R. éditeur.

Leichti, R.J. and Tang, R.C. (1989). Effect of creep on the reliability of sawn lumber and wood-composite I-beams. *Mathematical Computer Modelling*, **12**(2) 153–161.

Leivo, M. (1991). *On the stiffness changes in nail plate trusses.* VTT Publications 80, Espoo, Finland. 192p. + app. 46 p.

Littleford, T.W. (1966). *Performance of glued-laminated beams under prolonged loading.* Information Report VP-X-15. Department of Forestry and Rural Development, Forest Products Laboratory, Vancouver, B.C., Canada. p.21.

Martensson, A. (1990). Effect of moisture and mechanical loading on wood and wooden materials. IUFRO S5.02 Meeting, St John.

McNatt, D. and Superfesky, M.J. (1983). *Long-term load performance of hardboard I-beams*. Research Paper FPL441. USDA Forest Service, Forest Products Laboratory, Madison, WI, p.10.

Meierhofer, U. and Sell, J. (1979). Physikalische Vorgänge in wetterbeanspruchten Holzbauteilen. *Holz als Roh-und Werkstoff*, **37**(1979) 227–234.

Mohager, S. (1987). Studier av krypning hos trä KTH Stockholm, 139p.

Molinski, W. (1987). Deformations of wood stretched across the grain during simultaneous moistening and re-drying in humid air. *Holzforschung und Holzverwertung*. **39**(1987) 116–118.

Molinski, W. and Raczkowski, J. (1988). Creep of wood in bending and non-symmetrical moistening. *Holz als Roh-und Werkstoff*. **46**(1988) 457–460.

Molinski, W., Raczkowski, J. and Sarniak, J. (1987). Hydro-mechanical creep of wood in compression across the grain. *Holzforschung und Holzverwertung*, **39**(6) 148–151.

Mukudai, J. (1983). Evaluation of linear and non-linear viscoelastic bending loads of wood as a function of prescribed deflections. *Wood Sci. Technol*. **17**(1983) 203–216.

Nakada, O. (1960). Theory of non-linear responses. *J. Phys. Soc. Japan*, **15**.

Nakai, T. and Grossman, P.U.A. (1983). Deflection of wood under intermittent loading. Part I: Fortnightly cycles. *Wood Sci. Technol*. **17**(1983) 55–67.

Nielsen, A. (1972). *Rheology of building materials*. Document DG. National Swedish Institute for Building Research, Stockholm, Sweden. p. 225.

Palka, L.C. (1981). *Effect of load duration upon timber fasteners. A selective literature review*. Report to the Canadian Forestry Service. Forintek Canada Corp., Vancouver, B.C. p. 58.

Palka, L.C. (1989a). *Long-term strength of Canadian commercial waferboard: 5/8-inch thick panels in bending. Creep data and interpretation*. Annual Report to the Canadian Forestry Service. Forintek Canada Corp., Vancouver, B.C. p. 26 with Appendices. p.150.

Palka, L.C. (1989b). *Review of an exploratory study of truss-plate joints in tension under ambient laboratory conditions: I. Definitions, models and test methods. II. Traditional data interpretation. III. Reliability based design*. Annual Report to the Canadian Forestry Service. Forintek Canada Corp., Vancouver, B.C. p.85.

Palka, L.C. and Rovner, B. (1989). *Long-term strength of Canadian commercial waferboards: 5/8-inch thick panels in bending. Creep-rupture data and interpretation*. Annual Report to the Canadian Forestry Service. Forintek Canada Corp., Vancouver, B.C. p. 30 with Appendices p.12.

Palka, L.C. and Rovner, B. (1990). *Long-term strength of Canadian commercial waferboard 5/8-inch (16 mm) panels in bending. Short-term and long-term test data*. Forintek Canada Corp. Vancouver, B.C., Canada. 48 p. + Appendices (300pp.).

Pentoney, R.E. and Davidson, R.W. (1962). Rheology and the study of wood. *For. Prod. J*. **12**(5) 243–248.

Pienaar, F.R.P. and Brondhorst, B. (1982). *The creep deflection of South African pine structural timber in bending*. Special Report 254. National Timber Research Institute, Council for Scientific and Industrial Research, Pretoria, S.A. p. 9.

Pierce, C.B. and Dinwoodie, J.M. (1977). Creep in chipboard (Part I). *J. Materials Sci.*, **12**.

Pozgai, A. (1986). Time dependent and sorption deformation in calculation of wooden beam deflections. *Drevatsky Vyskum*. **109**(1986) 49–67.

Ranta-Maunus, A. (1975). The viscoelasticity of wood at varying moisture content. *Wood Sci. Technol*. **9**(1975) 189–205.

Ranta-Maunus, A. (1976). *A study of the creep of plywood*. VTT Structural Mechanics Laboratory; Report 5, Espoo, 85 p. + app.

Ranta-Maunus, A. (1991). Collection of creep data of timber. CIB-W18 Meeting, Oxford.

Rouger, F., Le Govic, C., Crubile, Ph., Soubret, R. and Paquet, J. (1990). Creep behaviour of french woods, *Timber Engineering Conference*, Vol. 2, pp. 330–336, Tokyo.

Sasaki, H. and Maku, T. (1963). *The creep of glued laminated wood beam*. Wood Research 31, Wood Research Institute, Kyoto University, Kyoto, Japan. p. 41–49.

Schaffer, E.L. (1972). Modelling the creep of wood in a changing moisture environment. *Wood and Fiber*. **3**.

Schapery, R.A. (1966). A theory of non-linear thermo-viscoelasticity based on irreversible thermodynamics. *Proc. Fifth US Nat. Congr. Appl. Mech.*, ASME.

Schniewind, A.P. (1966). On the influence of moisture content changes on the creep of beech wood perpendicular to grain including the effects of temperature and temperature changes. *Holz als Roh-und Werkstoff.* **24** (1966) 87–98.

Schniewind, A.P. (1968). Recent progress in the study of the rheology of wood. *Wood Sci. Technol.* **2**(1968) 188–206.

Schniewind, A.P. and Barrett, J.D. (1972). Wood as a linear orthotropic viscoelastic material. *Wood Sci. Technol.* **6**(1972) 43–57.

Schniewind, A.P. and Lyon, D.E. (1973). Further experiments on creep rupture life under cyclic environmental conditions. *Wood Fiber.* **4**(4) 334–341.

Senft, J.F. and Suddarth, S.K. (1971). An analysis of creep-inducing stresses in Sitka spruce. *Wood Fiber.* **2**(4) 321–327.

Simulski, S.J. (1989). Creep functions for wood composite materials. *Wood and Fiber Science.* **21**(1) 45–54.

Srpcic, J. and Houska, M. (1990). Creep factors for wood and glulam structures. *Proceedings of the 1990 International timber engineering conference*, p. 416–423.

Szabo, T. and Ifju, Y. (1970). Influence of stress on creep and moisture distribution in wooden beams under sorption conditions. *Wood Sci.* **2**(3) 159–167.

Toratti, T. (1991a). *Creep of wood in varying environment humidity.* TRT report 19, Helsinki University of Technology, Finland.

Toratti, T. (1991b). Long term bending creep of wood. CIB W18A, Oxford, 9 p.

Urakami, H. and Fukuyama, M. (1982). Application of rheological model to creep behaviour in the bending of wood and element constants. *Bull. Kyoto Univ. Forests.* **28**(7) 414–421.

Van Der Put, T.A.C.M. (1989). Deformation and damage processes in wood. Ph. D. Dissertation. Delft University Press, Delft, Netherlands. p.154.

Whale, L.R.J. (1988). Deformation characteristics of nailed or bolted timber joints subjected to irregular short or medium term lateral loading. Ph. D. Dissertation. South Bank Polytechnic, London, U.K. p. 260.

Wilkinson, T.L. (1984). *Long-time performance of trussed rafters with different connection systems.* Forest Products Laboratory. Madison, 19p.

Wong, P.C.K., Bach, L. and Cheng, J.J. (1988). *The flexural creep behaviour of OSB stressed skin panels.* Structural Engineering Report 158, Dept. of Civil Engineering, University of Alberta, Edmonton, AL. p.138.

Ylinen, A. (1965). Prediction of the time-dependent elastic and strength properties of wood by the aid of a general non-linear viscoelastic rheological model. *Holz als Roh-und Werkstoff.* **23**(7) 193–196.

Zaupa, F. (1989). Time-dependent behaviour of glued-laminated beams. Paper at CIB W18/IUFRO S5.02 Joint Meeting, Florence, Italy. p.21.

3

Sensitivity of creep to different constant environments

C. Le Govic

3.1 Notation

t : time of observation
t_0: reference time corresponding to the determination of elasticity
T : temperature
w : moisture content
J : compliance
$\theta(t,T,w)$: creep factor or reduced creep compliance
$$\theta(t,T,w) = J(t,T,w) / J(t_0,T,w) \tag{3.1}$$
$\Phi(t,T,w)$: relative creep factor
$$\Phi(t,T,w) = J(t,T,w) - J(t_0,T,w) / J(t_0,T,w) \tag{3.2}$$

Directions of anisotropy labels are:

- Radial: 1
- Tangential: 2
- Longitudinal: 3

3.2 Introduction

Builders have always been aware of the importance of creep in materials to the long-term behaviour of structures and have considered the matter either in terms of deflection or long-term strength. Wooden structures are subject to two kinds of different creep:

- creep in members which is solely dependent upon the long term behaviour of wood;
- creep in joints which is more complex as it includes phenomena linked to the characteristics of connectors (often metallic) and to their interfaces to wood (damage, etc.).

With regard to all the of phenomena involved, this chapter describes the delayed behaviour of wood including its anisotropy and the temperature and moisture content influence, related to the modelling of the observed behaviours. Our bibliographic documentation only recapitulates the articles written after 1960.

From a rheologic point of view, creep in wood is a feature of its visco-elastic behaviour that is linked to its polymeric nature. To go further into the main subject of this chapter, we will present some concepts about the viscoelastic properties of polymers. Our references on this rheological background are Huet 1988 and Le Govic (1988).

3.3   Viscoelasticity of solid polymers

A viscoelastic material can characterize with creep-recovery tests and relaxation experiments.

3.3.1   Isothermal creep, relaxation and recovery.

Creep is the increase of strain under constant stress. When suppressing this stress, strain goes on decreasing, this phenomenon being partial or total. These two phenomena are particular representations of the viscoelastic (and/or plastic) behaviour of wood. Figure 3.1 gives an illustration of these different behaviours.

The plastic component is discriminating with relaxation and shape restoring test. In a creep test the non-recoverable character of strain is not a criterion for the statement of a plastic component. For wood, in this way, we did not find out any article on these discriminent kinds of tests.

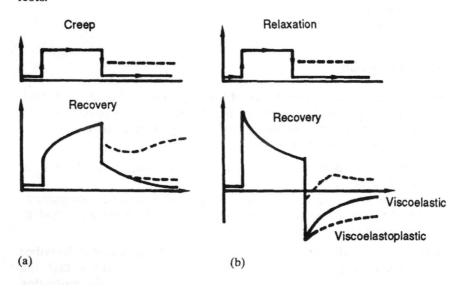

Fig. 3.1 Reponse of a viscoelastic material to a:
   a) creep-recovery test
   b) relaxation-recovery test.

3.3.2 Linear viscoelasticity (under constant climate)

In a general case, the response of a viscoelastic material (in strain for creep) is a functional depending upon prior history of sollicitations. In this case, the material is called hereditary.

Viscoelasticity behaviour can be linear, which is a result of the Boltzmann superposition principle. In this case, the functional can be distinguished whether the response appears as strain or stress. Then the strain under variable stress can be represented by linear integrals with convolutive Kernel:

$$\in (t) = \int_0^t S(t - u) . d\sigma(u)$$

$$(3.3)$$

where u is the time corresponding to the application of the increment of stress $d\sigma(u)$,

σ stress and $\in$ strain are tensors of second order

S (creep or compliance function) depending only of (t-u) in the non-ageing case.

For a stress history where $\sigma_o(t) = H(t).\sigma_o$, H(t) being the Heaviside function, creep function in a tridimensional case expresses itself under a tensorial form:

$$\in (t, \sigma_0) = S(t) . \sigma_0$$

$$(3.4)$$

The number of independent components is more or less reduced according the case of symmetry.

In a unidimensional case, the linear behaviour links strain to the constant stress by a function depending only upon climate and environmental static conditions.

3.3.3 Principle of time–temperature equivalency

Terms of the principle
Wood is a material made of polymers. Generally, experimental observation of polymers shows the existence of :

1) a vitreous state corresponding to the elastic behaviour resulting insensitive neither to the temperature nor to the time of observation (frequency).

The 'real elasticity' appears insensitive to the temperature (with respect to time). Therefore, concerning wood, the different data relative to the dependence of elasticity modulus with temperature is, (Gerhards 1982), correspond in fact to the evolution (of the contrary) of viscoelastic compliance which is generally determined between 30 seconds and 5 minutes.

2) different transitions (second order, vitreous) where behaviours are

sensitive to temperature and to time of observation (frequency). A transition means that some dissipate phenomena may occur (in a defined time–temperature range) between two situations where they do not occur.

Concerning synthetic polymers, the thermal dependence to temperature could be particular and expresses itself under the time-temperature equivalency principle. The use of the time–temperature principle (which is in fact an assumption which must be experimentally verified) means that in a defined time–temperature range, where the transition takes place, time flow and temperature have the same effect on the molecular mechanisms responsible for the mechanical behaviour: the temperature is an accelerator of time flow.

This property leads to the introduction of a reduced variable. It allows the expression of creep function not only from the triple dependency of time (t), temperature (T) and humidity (w) but also under the dependency of a reduced variable $t/\tau(T;w)$ and w. The variable (τ) depends only upon temperature and moisture content. For two different temperatures, $T_1 T_2$, the time–temperature principle can be written:

$$t_1/t_2 = (T_1)/\tau(T_2) = a(T_2) \tag{3.5}$$

A main value is the shift factor $a(T_2)$, where T_1 represents the reference temperature which leads to the notion of master curve well known to the polymerists.

The logarithm plot of the shift factor with regard to the inverse of the absolute temperature allows us to understand the nature of the transition (vitreous, secondary order transition) (Le Govic et al. 1989).

The dependency of τ with temperature can be written under Arrhenius' Law (3.5) in the case of a secondary transition and under the Tammann and Hesse law (3.7), in the case of a vitreous transition. WLF law being a particular of Equation (3.7).

$$\tau(T) = \tau_0 \exp(W_0/RT) \tag{3.6}$$

$$\tau(T) = \tau_0 \exp[\,W(T)/R(T-T\infty)\,] \tag{3.7}$$

where τ_0 is a constant, referenced to infinite time,
R is the constant of perfect gas,
W is the activation energy (constant for an Arrhenius' law),
T∞ is temperature to which free volume is zero.

In comparison with the deformation kinetic model (the Eyring's model) the activation energy is to be considered as a useful empirical parameter.

Complex plane representation
In experimental facts, the experiments made in order to settle this principle are of vibratory types, the approach being identical to the complex modulus. Representation in the complex plane where the imaginary part of the complex modulus is versus the real part of it, allows us to have a clear idea of the validity of this principle.

When this principle is satisfied, all isotherms create one curve like a direct master curve. Complex modulus E(ir, T;w) depends on this precise case of the two variables i r (T,w) and of w.

Several authors have proposed methods for giving a representation of creep in the complex area. Rouger chose the Alfrey method where complex compliance

$A^* = A_r + i A_i$ is approximated by:

$$A_r = A \qquad\qquad (3.8.a)$$

$$A_i = /2 * \partial A / \partial \text{Ln } t \qquad\qquad (3.8.b)$$

A being the creep compliance.

3.3.4 The non-linear viscoelasticity

Some theoretical developments are existing in non-linear viscoelasticity. This theory is well covered by Lockett (1972) and Schapery.

Very little applications of it, exist for wood: Huet (1984) explained the potential of the multiple integrals (Frechet theory) to model the non-linear viscoelastic behaviour of wood, step undertaken by Ranta-Maunus (1972) and developed to take into account the mechano-sorptive effects. Van Der Put (1972): has developed the Eyring model which includes the effects of environment changes (mechano-sorptive effect).

3.4 Experimental phenomena

We shall summarize here the best know experimental facts on wood creep and shall give more general data relevant to the engineer.

The evolution of the creep strain with respect to time largely shows two kinds of characteristics (Fig. 3.2).

Creep is (very) fast at the beginning and (sometimes) difficult to distinguish from instantaneous behaviour; this is the primary creep which corresponds to a non-constant speed of creep i.

Creep then slows down very strongly and seems to stabilize quickly; this is the secondary creep where the rate of creep is constant.

The third kind of creep which happens after the two previous ones is

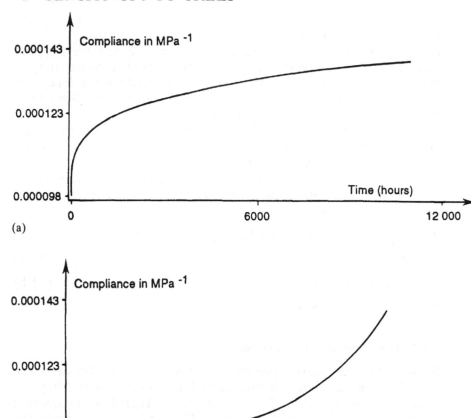

Fig. 3.2 Typical creep curves:
   a) in a linear time scale
   b) in a logarithmic time-scale.

studied in terms of duration of load and not in the continuity of the two others; tertiary creep is an accelerated creep in arithmetic time plot. Representation of creep strain (or compliance) on a time logarithmic scale allows us to show that stabilization of creep is only apparent as in this representation logarithmic speed is growing up.

This means that no finite creep limit is seen at realistic times with the use of logarithm time plot. With the use of the logarithm representation we can observe that the slow slope(s) appears before a larger one which corresponds to the instantaneous behaviour (less than 30 seconds).

3.4.1 Limits of linear viscoelastic behaviour

In the case of a well-defined linear relation between stress and strain independent of time, the creep function allows, for a moderate programme of sollicitations, under constant climate the calculation of the strain evolution.

In practice, experimental results would determine the levels of stress (expressed in percentage of strength) related to the nature of sollicitations and their orientations, regarding the anisotropy axes, environmental conditions, species, etc. Table 3.1 gives a bibliographic synthesis.

Generally, experimental results only deal with the sole longitudinal direction and correspond to the data of an average value without any complementary statistic indications (standard deviation). Only two authors mention these kinds of data (Miller and George (1974) and Le Govic *et al.* 1990).

The reading of table 3.1 seems to indicate that the limit of linearity in bending (expressed as a percentage of strength) decreases with moisture content. This stress limit is quite low (30–35 %, 18–21 MPa). The one in tension seems higher (nearly 50%) whereas the other in compression is difficult to determine reading Table 3.1.

Table 3.1 Limits in stress of linear viscoelastic behaviour. (Level of stress is expressed as a percentage of strength corresponding to the same conditions).

Authors	Species	Kind of stress	Load duration	Stress level	Environmental conditions
KINGSTON	Pinus L	Bending	17 hours	40 %	11%, 21°C
LE GOVIC	Spruce L	Bending	1 week	30 %	18 %, 25°C
MILLER	Spruce L	Bending	30 days	40 %	12 %
MUKUDAI	Hinoki L	Bending	10 hours	35 – 45 %	11 %, 25°C
DINWOODIE	Spruce L	Compression	–	25 %	12 %, green
KEITH	Spruce L	Compression	1 day	57 – 70 %	9 % – 18 %
KING	Basswood L	Tension	30 min	30 – 40 %	12 %
BHATNAGAR	Teak L	Tension	5 hours	50 %	12 %, 30°C
HAYASHI	Spruce L	Tension	4 days	60 %	12 %, 20°C
FOUDJET	Tropical hardwood L	Isostress bending	2 weeks	35 %	18 %
YOUNGS	Oak Transverse	Tension	70 days	40 %	12 %, 27°C

3.4.2 Anisotropy

The long term behaviour of wood exhibits a drastic anisotropy and is characterized by viscoelastic compliance which are linearly independent S_{ijkl}:

$$\epsilon_{ij}\,(t,\sigma) = S_{ijkl}\,(t)..\,\sigma_{kl} \qquad \text{in 4 subscripts notation}$$

$$\epsilon_{\alpha}\,(t,\sigma) = S_{\alpha\beta}\,(t).\,\sigma_{\beta} \qquad \text{in a matrix representation}$$

Using the same symmetrical considerations as for the elastic case, the matrix numbers of compliance are assumed to be nine. To completely establish this matrix, there exists a certain number of experimental problems. Schniewind (1968) and Cariou (1987) are the most exhaustive on this matter. Studies by Hayashi (1993) are of a more methodological level.

The experimental establishment of viscoelastic compliance is reduced to a bidimensional problem. Two kinds of experimental approaches exist:

- Single tests (in terms of stress): experiments are done in the anisotropy directions or from remade specimens, for shear tests: this is the approach of Cariou (1987).
- Out off (anisotropy) axes tests, which correspond to bi-axial (stress) experiments which is the approach of the two other authors.

Schniewind and Barrett (1972) have determined viscoelastic compliance by out off anisotropy axes, without using the adequate methodology as for the simultaneous determination of the two normal and shear components, which is the case of Hayashi. These authors have pointed out that creep related to 1000 minutes in shear (term S_{44}) corresponds to 60% of the one in the tangential direction. For the same time of reference, the creep factor for longitudinal is 5 times smaller than the one in shear.

Cariou (1987) has determined on the same species for three moisture content, the whole of the nine viscoelastic compliance for a duration of load of 1400 minutes. This experimental work is completed by a power law modelling. In addition, this author has demonstrated that the terms $S_{iijj}(t)/S_{iiii}(t)$ for the subscripts (i,j = 1,2,3;i j), which can be assimilated to Poisson's ratios, were not varying a lot in time. The two transversal values were not studied because the measurements were too difficult. Schniewind found an evolution of these values ranking from 10% for a creep duration of 1000 minutes.

We notice that we have not found any systematic investigation on creep behaviour in tension and compression whereas behaviour in bending which has been studied more, combines compression and tension.

3.4.3 Influence of moisture content

Creep in wood increases with moisture and this in spite of the elastic softening with moisture content, (Gerhards 1982).

Schniewind (1968) reports experimental works concerning different types of solicitations. Bach and Pentoney (1968) mention a variation of the viscoelastic compliance with the square of moisture content.

Table 3.2, taken from Cariou (1987) gives the creep factors corresponding to the nine compliance according to three moisture contents.

Table 3.2 Instantaneous value corresponds to one minute creep. Values in brackets correspond to the standard deviation.

COMPLIANCE $1 + k_{creep}$ (one day)	\	MOISTURE CONTENT	\
	7%–9%	10%–12%	18%–20%
S11	1.29	1.35	1.56
	(5.2%)	(6%)	(4.7%)
S22	1.25	1.34	1.54
	(3.6%)	(2.3%)	(5.5%)
S33	1.076	1.09	1.107
	(1.4%)	(3.8%)	(3%)
S44	1.09	1.13	1.18
S55	1.127	1.15	1.14
S66	1.17	1.23	1.265

Creep of wood is more important in tension perpendicular to the grain than in shear which appears, in fact, less sensitive to moisture content.

3.4.4 Influence of temperature

Different authors have studied the direct effect of temperature on the increase of creep: Davidson (1962) and Le Govic et al. (1988) in bending parallel to the grain (Fig. 3.3); Bach and Pentoney (1968) in longitudinal tension reporting an evolution of compliance with the square of temperature in the range of 30° and 70°; Kingston and Clarke (1962) in compression parallel to the grain; and Youngs (1957)in transverse tensile creep.

The time–temperature equivalency principle is invalidated for time and temperature ranges corresponding to Davidson's works (1962) and for the Kauman study (Kauman et al. 1987) held in a more complex representation which shows the existence of two transitions (Fig. 3.4) corresponding to the observation of stabilization of creep (55°C, 2 days).

Viscoelastic data obtained from vibratory experiments prove the existence of several transitions (Le Govic et al. 1988). The principle of time–temperature equivalency does not apply to the whole of the

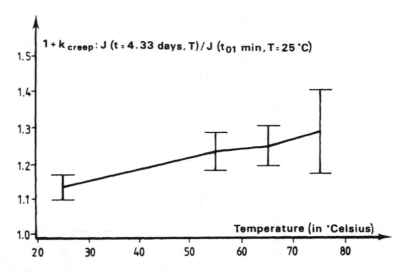

Fig. 3.3 Evolution of creep factor with temperature. Longitudinal bending, 12 % MC.

manifestation of viscoelastic behaviour but to the inside of each transition.

Moreover temperature sensitivity of 'standard' elasticity demands that the creep factor has to be accompanied by a definition of the elastic measurement.

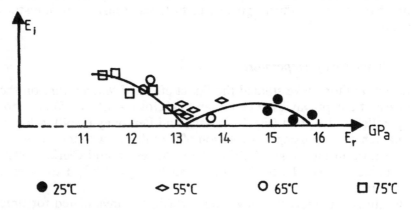

Fig. 3.4 Representation in the complex plane of creep data (by Alfrey approximation).

3.5 Linear creep models

The proposed linear models (compliance is independent of stress) can be divided into two principal groups:

- rheological analogic models made from spring and dashpots leading to exponential models;
- rheological models made from a combination of spring(s) and parabolic element(s) leading to power law kinetics.

To these, we must add logarithmic models already developed in the area of non-linear viscoelasticity.

In a general way serial set up models are more easily identified from creep experiments (or complex modulus results). Moreover serial assembly and parallel assemblies are fully equivalent, as is well known, for classical spring and dashpot models and for power law units (Huet 1984).

From a heterogeneous material point of view, half of the stress components are continuous across an interface, and for the other half it is the strain which is continuous. Thus, there is no physical reason to give a preference to the assembly of kinetic units, so we insist on serial assembly, having in mind that creep models must have a sufficient fit of relaxation experiments (and dynamic tests).

3.5.1 Power law model

The most used model with regard to experimental creep of wood is the power law model (see Nielsen 1984). It is widely used in the study of other materials (polymers, concrete, etc.) and seems to give a good description of the behaviour described in Section 3.3 (small time–temperature range).

The relaxation function as given by Eq. (3.9.c) does not fit the experimental behaviour of wood well because of the decreasing to zero of stress in the relaxation test. Moreover such a model cannot fit the vibratory experiments. The mathematical form of this model differs with authors: Youngs, Ota, Schniewind and Barrett (1972), Le Govic *et al.* (1988). It is pertinently written in terms of compliance by the following equations:

$$A(t) = A_0 (1+(t/\tau^k) \tag{3.9a}$$
$$S_{\alpha\beta}(t) = S_{\alpha\beta}(1+(t/\tau)^k) \tag{3.9b}$$

where τ is the doubling time of elasticity: $A(\tau) = 2A_0$,
$A_0 S_{\alpha\beta}$, is the elastic compliance,
k is the kinetic creep factor $0 < k < 1$.

$$r(t) = \sum\nolimits_{n=0}^{\infty} (-X)^n / \Gamma(1+nk) \quad \text{with } X = (1+k)\,(t/\tau)^k \tag{3.9c}$$

where the summation term is recognized as being the Mittag–Leffler function, and $\Gamma(x)$ the second Euler function. For $k < 1/3$ (which is the case with wood) we have a very simple approximation given in (3.9d).

$$r(t) = [1 + (t/\tau)^k]^{-1} \qquad (3.9d)$$

The advantages of this creep function are two fold:

- There are a low number of parameters with equal sensitivity to be determined.
- A simple integration of thermoactivation law on the variable τ, t/τ have been the properties of a reduced variable.

All the parameters are very sensitive to the choice of elasticity: a difference of 5% on this leads to a difference which can reach 150% upon the other parameters (Rouger1988). All the different studies, except the CTBA's work, determine elasticity equal to standard compliance.

A range of power law parameters corresponding to the whole of viscoelastic compliance is reported in Table 3.3. This table is presented as a comparison with data from studies made at the CTBA and bibliographic data reported by Nielsen (1984) which are a bit different.

Table 3.3 Power law parameters corresponding to the nine viscoelastic compliances

Author	Temp. MC	Test	Compliance	k	minutes $\tau(T)$	S° MPa^{-1}
CARIOU	25°C, 12.4 %.	Tension RT	$S_{11}(t)$	.28	1 E5	0.939 E-3
CARIOU	25°C, 12.7%	Tension TL	$S_{22}(t)$	.16	2.6 E6	1.362 E-3
CARIOU	25°C, 13.7%	Tension TR	$S_{22}(t)$	.22	13 E6	1.427 E-3
CARIOU	25°C, 12.5%	Tension LT,LR	$S_{33}(t)$	.16	(4.8-6.9) E10	(9.69-8.07) E-5
CARIOU	25°C, 13.3%	Shear LT	$S_{44}(t)$	.15	5.5 E8	.5778 E-3
CARIOU	25°C, 12.7%	Shear LR	$S_{55}(t)$	.25	3.5 E6	.7381 E-3
CARIOU	25°C, 11.8%	Shear RT.	$S_{66}(t)$	.2	8.9 E5	4.734 E-3
CARIOU	25°C, 14%	Tension LR	$S_{31}(t)$	1.7	1 E12	2.64 E-5
CARIOU	25°C, 12.5%	Tension LT	$S_{23}(t)$	.15	1.9 E10	5.075 E-5
CARIOU	25°C, 13.7%	Tension TR	$S_{21}(t)$	.18	1.4 E6	0.603 E-3
CARIOU	25°C, 12.4%	Tension RT	$S_{21}(t)$	.25	0.11 E6	0.539 E-3
HAYASHI	25°C, 12%	'Out off' Tension	$S_{11}(t)$	.214	2.2 E5	0.552 E-3
HAYASHI	25°C, 12%	'Out off' Tension	$S_{13}(t)$	.127	7.3 E7	2.47 E-5
HAYASHI	25°C, 12%	Tension	$S_{55}(t)$	.23	9.4 E5	0.212 E-3
HAYASHI	25°C, 12%	Tension LR	$S_{55}(t)$	.27	5.9 E5	0.743 E-3
HAYASHI	20°C, 12%	Tension L	$S_{33}(t)$	.135	1.4 E12	6.4 E-5
LE GOVIC	25°C, 12%	Bending L*	$S_{33}(t)$	.11	6.3 E10	5.87 E-5
NIELSEN	20°C, 15%	Bibliographic	$S_{33}(t)$	.2 –.25	1.5 (E7 to E8)	–

*Tests including different temperatures

3.5.2 Exponential models

The principal form is made of a serial assembly of Kelvin elements in association or not with Maxwell components (viscous term):

$$J(t) = J_0 + \sum_{i=1}^{i=n} (1 - \exp(t / \tau_i)) + \underbrace{t / \tau_0}_{\text{viscous term}}$$

(3.10)

The Burger-model which is a Kelvin model in serial assembly with a Maxwell model leads to a relatively good fit of both creep and relaxation experiments in small time–temperature range but the complex image in Cole–Cole representation (which is half a circle) does not correspond to experimental behaviour.

The time response of such functions $[f(t) = 1\text{-}\exp(\text{-}t)]$ is one decade. Therefore, to be realistic with regard to experimental data, the number of Kelvin components is variable according to creep duration.

Mukudai (1987) has used a standard model of Kelvin (cf Eq. 3.10) for the term $S_{33}(t)$ in the bending case and $S_{55}(t)$ with the number of links equal to three. The maximum proposed number of elements was 40 for Davidson (1962) in longitudinal bending for experiments including different temperatures.

Different authors have studied these exponential models (Gressel 1984; Grossman and Davidson 1962). It corresponds very little to experimental facts (Senft and Suddert 1971). Other exponential models have been studied (Burgers and Ylinen 1965).

All the interest of this model remains in the exponential function which allows recurrency relations for a discretization in time to an eventual application in computing values for timber structure codes.

3.6 Thermoactivated models

To take account of both the time and temperature dependence of creep, different authors have proposed linear viscoelastic models including time and temperature as variables.

3.6.1 Thermoactived power law model

This model combines a 'power law' corresponding to the isothermal creep behaviour (3.9a) and a thermoactivation law (Arrhenius' case (3.6)), which leads to a linear equation with 4 parameters and two variables (3.11). This model includes time and temperature (static value) dependence of wood creep behaviour.

$$A(t) = A_0 (1 + (t/\tau_0)^k) \exp(\text{-}kW_0/RT)$$

(3.11)

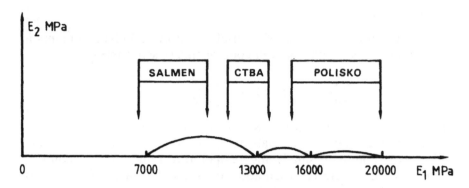

Fig. 3.5 'Parabolic multi-transitions model' qualitatively fitted to three transitions identified at this time.

with: $A_0 = 0.5866 \ 10^{-4} \ \text{MPa}^{-1}$, $(E_{standard} = 16 \ 400 \ \text{MPa})$
        $W = 162.66 \ \text{kJ/mole}$
        $k = 0.112$
        $\tau_0 = 1.22 \ 10^{-16} \ \text{s}$

This model has been proposed for the fitting of creep experiments for different temperatures, (Kauman *et al.* 1987).

The establishment of such a model depends on the approximation of the existence of a master curve (one transition) justifying itself to its weak intensity of the first transition (Fig. 3.5) and the good restitution of experimental creep data.

This model has been used in order to predict the creep factor relative to the different Eurocode 5 duration of load classes (Le Govic *et al.* 1988) at 12% MC.

3.6.2   Complex model with parabolic kinetic

C. Huet (1984) has proposed a model built of the obvious evidence of several transitions for wood. This model is not still in a position to fit quantitatively the viscoelastic behaviour of wood but allows us to describe this behaviour as a whole in a wide range of time–temperature (see Fig. 3.4).

This model is a serial link of parabolic elements, Cole–Cole solid, (3.10.b), each link being made of a spring and a parabolic component in parallel. In this case, compliance is a complex term (3.10.b). Each parabolic component is submitted to time–temperature equivalence written in terms of the Tamman and Hesse law (3.7). The corresponding complex modulus is given by (3.12.c).

$$A^*(i, T) = A_{00} + \sum_{.\alpha=1}^{.n} A\alpha(i \quad \tau_\alpha)$$

$$\tag{3.12.a}$$

$$A^*(i,T) = A_{0\alpha} [\ 1+ (i\ \tau_\alpha)^{k\alpha}] \tag{3.12.b}$$

$$E^*(i,T) = 1 / A^*(i ,T) \tag{3.12.c}$$

3.7 Representation of creep by non-linear models

Several non-linear models by development of linear models have been put forward:

- logarithmic model   Bach and Pentoney (1968)
- power law model   Kitahara and Okabe (1959)
- Burger's model   Ylinen (1965)
- Kelvin generalized model   Mukudai (1987)

Rantas-Maunus (1972) has proposed an original approach in the field of wood (plywood) by the use of the multiple integrals method developed after for mechano-sorptive effect. To make obvious the behaviour of birch plywood, the following equation has been proposed:

$$E(t)/E(0)= 1 + 0.094\ t^{0.2} + 31\ w^3\ \sigma t^{0.5} \tag{3.13}$$

Van Der Put (1989) has developed a non-linear model based on the Eyring transport theory linked with the kinetics of the bond breaking processes. It will be noticed that the effect of the stress depends on its nature (tension or shear or compression). For polymers the Eyring model is the most satisfactory for the shear component. An interesting summary is done in Huet (1990).

3.8 Conclusions

General reading of worldwide scientific literature relative to creep in wood turns up the four following points.
1) Some studies have been made on the subject but are not accessible because their publication is not in English: France, Japan, Scandinavian countries and several Eastern countries where no references could be found.
2) Published literature is very fragmentary from the point of view of results and general conclusions on the following points.
   - Transverse behaviour, shear and compression sollicitations have been subject to very little important experimental investigation corresponding to 'creep in wooden structures'.

- If important variables exercising influence on creep (temperature, moisture, anisotropy) have been studied, the conclusions are
  limited either by an insufficient modelling or by the analysis
  resulting from it. Very few studies refer to the linearity of
  viscoelasticity (establishment of the linear domain, recovery, and
  different kinds of load,) which has a negative impact on modelling.

3) Creep modelling by 'power law' is the most suitable to fit the experimental behaviour of wood studied in research laboratories. Up to
  now, we have a complete range of parameters for the nine viscoelastic compliances in wood at constant climate (different moisture
  contents), but this type of law is hardly integrable to calculation
  codes and modelling by exponential law keeps all its meaning. From
  this point of view it would be interesting if researchers modelled the
  observed behaviours with these two kinds of law.

4) It seems that we could predict long-term creep behaviour by
  studying short-term behaviour thermically activated thanks to a
  well-controlled application of the time–temperature equivalency
  principle.

3.9 References

Bach, L. and Pentoney, R.E. (1968). Non linear mechanical behaviour of wood. *For Prod. J.*, **18**(3), 60–66.

Bhatnagar, N.S. (1964). Creep in wood in tension parallel to grain. *Holz Als Roh-und Werkstoff*, **22**(8), 296–299.

Cariou, J.L. (1987). Caractérisation d'un matériau viscoélastique anisotrope, le bois. Doctoral thesis, University of Bordeaux, France.

Davidson, R.W. (1962). The influence of temperature on creep in wood. *For. Prod. J.*, **12**(8), 337–381.

Dinwoodie, J.M. (1981). Timber its nature and behaviour, Van Nostrand, London.

Fondjet, A (1986). Contribution à l'étude rhéologique du matériau bois, Doctoral Thesis, University of Lyon, France.

Gerhards, C.C (1982). Effect of moisture content and temperature on the mechanical properties of wood: an analysis of immediate effects. *Wood and Fiber*, **14**(1), 4–36.

Gressel, P. (1984). Prediction of long term deformation behaviour from short term creep experiments. *Holz Als Roh-und Werkstoff*, **42**(8), 293–301.

Hayashi, K., Felix, B. and Le Govic, C. (submitted). Wood compliances determination with special attention to measurement problems.

Holzer, S.M., Loferski J.R. and Dillard, D.A. (1989). A review of creep in wood: concepts relevant to develop long term behaviour predictions for wood structures. *Wood and Fiber Science*, **21**(4), 376–392.

Huet, C. (1984). Communication to the CEE seminar on wood.

Huet, C. (1988). Modelizing the kinetics of thermo-hygro-viscoelastic behaviour of wood in constant climatic conditions. *Proceedings of the 1988 Inter. Conf. on Timber Engineering*, Seattle, USA.

Huet, C.(1990). *Creep and time to failure in wood and wood structures*. Subproject III, wood material and wood products, Contract MA1B - 0049 - (FEDB). Final report to CEE.

Hayashi, K., Felix, B. and Le Govic, C. (1993) Wood viscoelastic compliances determination with special attention to measurement problems. *Materials and Structures*, **26**(160) 370–376.

Kauman, W.G., Huet, C., Felix, B., Hadjhamou, A., Le Govic, C., and Rouger, F. (1987). Recherches en rhéologie, rapport de fin d'étude. décision d'aide 85.G.0185.01 au Ministère de la Recherche et de l'Enseignement Supérieur, CTBA, Paris. S.S.

Keith, C.T. (1972) The mechanical behaviour of wood in longitudinal compression. *Wood Sci.* **4**, 234–244.

King, E.J. Jr. (1961). Time-dependent strain behaviour of wood in tension parallel to grain. *For. Prod. J*, **11**(3) 156–165.

Kelley, S.S., Rials, T. and Glassez, G.(1987). Relaxation behaviour of the amorphous components of wood. *J. of Nat. Sciences*. **22**, 617–624.

Kingston, R.S.T. and Clarke, L.N. (1962). Some aspects of the rheological behaviour of wood; I the effect of stress with particular reference to creep. *Aust. J. Appl. Sci*, **12**(2), 211–227.

Kitahara, K. and Okabe, N. (1959). The influence of temperature on creep of wood by bending test. Faculty of Agriculture, University of Tokyo.

Le Govic, C.(1988). Le comportement viscoélastique du matériau bois en relation avec sa constitution polymérique. Rapport bibliographique, CTBA , Paris.

Le Govic, C. (1991). Modeling wood linear viscoelastic behaviour in constant climate, Workshop 'fundamental aspects on creep in wood', COST 508, March 20–21 Lund 1991.

Le Govic, C., Hadj Hamou, A., Rouger, F. and Felix, B. (1988). Modélisation du fluage du bois sur la base d'une équivalence temps-température. *Actes du 2ème Colloque Sciences et Industries du bois*, Nancy (avril 1987), A.R.B.O.L.O.R éditeur, Nancy, France, pp. 349–356.

Le Govic, C., Benzaim, A. and Rouger, F. (1989). Creep behaviour of wood as a function of temperature: Experimental study, modelling and consequences for design codes. *Second Pacific Timber Engineering conference*, Vol 2, pp. 273–279, Auckland, New Zealand.

Le Govic, C., Benzaim, A. and Rouger, F. (1990). Recherches en rhéologie sur le matériau bois: 'Le comportement viscoélastique du bois – influence de l'humidité et de la température', rapport de fin d'étude. décision d'aide 87.G.0337, au Ministère de la Recherche et de l'Enseignement Supérieur, CTBA, Paris.

Lockett F.J. (1972). *Non linear viscoelastic solids*. Academic Press, London.

Miller, D.G. and George, P. (1974). Effect of stress level on the creep of eastern spruce in bending. *Wood Sci*, **7**(1) 21–24.

Mukudai, J. (1983). Evaluation of non linear viscoelastic bending deflection of wood. *Wood Science and Tech*. **21**, 39–54.

Mukudai, J. (1987). Evaluation of linear and non linear viscoelastic bending loads of wood as a function of prescribed deflections. *Wood Science and Tech*. **17**, 203–216.

Nielsen, L.F. (1984). Power law creep as related to relaxation, elasticity, damping, rheological spectra and creep recovery, with special reference to wood IUFRO; Timber Engineering Group meeting, Xalapa, Mexico.

Norimoto, M., Gril, J., Minato, K., Okurama, K., Mukadai, V. and Rowell, R. M. (1987). Suppression of creep of wood under humidity charge through chemical modification. *Mokusa Kogyo*, **42** (II), 504–5.

Polisko, V (1986). Anisotropy of dynamic wood viscoelasticity, *Nineteenth Colloquium of the Groupe Français de Rhéologie*, Cepadues Ed, Toulouse, France, pp. 453–460.

Ranta-Maunus, A. (1972). *Viscoelasticity of plywood under constant climatic conditions*. Technical Research Center of Finland Report, Helsinki, Finland.

Salmen, I. (1984). Viscoelastic properties of in situ lignin under water satured conditions. *J. Mat. Sci*. **19**, 3090–3096.

Schniewind, A.P. (1968). On the influence of moisture content changes on the creep of beech wood perpendicular to the grain. *Holz als Roh–und Werkstoff* **24**(3), 87–98.

Schniewind, A.P. and Barrett, J.D. (1972). Wood as a linear orthotropic viscoelastic material. *Wood Science and Tech*. **6**, 43–57.

Senft, J.F. and Suddart, S.K. (1971). An analysis of creep inducing stresses in sitka spruce. *Wood and Fiber* **2**(4), 321–327.

Van Der Put, T.A.C.M. (1989). *Deformation and damage processes in wood*. Delft University Press, Delft, The Netherlands.

Ylinen, V. (1965). Prediction of the time dependent elasticity and strength of wood by the aid of a general non linear viscoelastic rheological model. *Holz Als Roh-und Werkstoff*, **23**(7), 193–196.

Youngs R.L. (1957). The perpendicular to grain mechanical properties of red oak as related to temperature, moisture content and time, F.P.L. Report 2079.

4
Deformation kinetics

T.A.C.M. van Der Put

4.1 Introduction

Because it becomes increasingly necessary to be able to give a precise prediction of time-dependent behaviour in all circumstances it becomes necessary to follow the appropriate physical theory (statistical mechanics), deformation kinetics, that applies for all materials and explains the temperature dependence and all other aspects of the essential non-linear behaviour of materials.

The statistical mechanical theory of rate processes deals with all changes at the molecular level including those occurring in the deformation of solids. If there is a chemical potential, not only are the common chemical reactions explained but also, for example, the rate equations of the theory of diffusion. If the potential is electrical in nature, the theories of, for example, conduction, dielectric displacement and over-voltage are provided. If the potential is applied by a stress field, the theories of viscous flow or plastic deformation and fracture, etc. are obtained. Not only are these transport processes interrelated, but several of them may be combined. For instance, diffusion is necessary for the movement of dislocation jogs and also mechano-sorptive slip only occurs at moisture content changes. Often only the different constraints are determining as for instance for self-diffusion under the effect of a stress field or for the Herring–Nabarro mechanism (vacancy diffusion) of plastic deformation, or for interstitial diffusion under the effect of stress, where the activation energies of diffusion and of creep are the same. Knowing for instance the activation energy and temperature for melting, the parameters of the kinetic response in all other cases where the mechanism is operative, such as creep, fracture and self-diffusion, are known. Thus deformation is just one aspect of these known processes treated by deformation kinetics and the activation parameters are the same as those of the other aspects (dielectric, paramagnetic, diffusion, free volume, etc., constants).

Deformation kinetics theory thus explains transport phenomena of deformation such as creep, plasticity, fracture, dislocation formation and propagation, diffusion, crystallization, glass transition, rubber behaviour, annealing, ageing, etc. The theory is not just a model comparable with the other applied phenomenological models but is the only way to describe physical reality. The dislocations and their rate of propagation with the caused rate of the local plastic deformation, can be seen directly by different methods such as etching, birefringence, X-ray diffraction, transmission electron microscopy and field ion microscopy, giving a direct verification of the aspects of statistical mechanics and the relation to macro-deformation.

An example of this possibility is given by the annealing of glass (Van Der Put 1991) (that is the same as for organic glasses as in wood). The

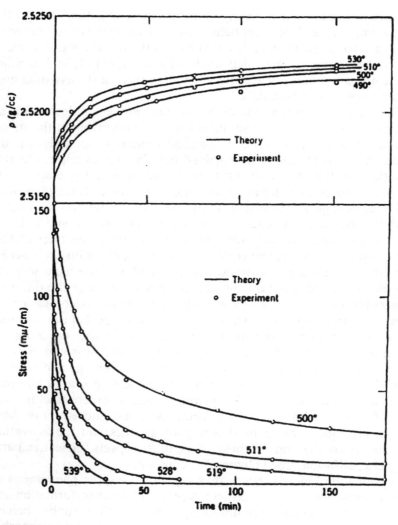

Fig. 4.1 Stress relaxation and volume contraction during annealing of glass.

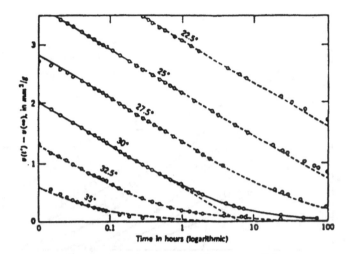

Fig. 4.2 Volume contraction of a wood cellulose polymer after cooling from well above transition temperature to the temperatures indicated.

birefringence (a kind of photo-elastic effect) shows the 'dislocations' and the stress field around them (and even the magnitude of the stress). Now the internal stress relaxation or change of birefringence and the decrease of volume during annealing follow exactly the theory of Van Der Put (1989) where the same equation can be written in terms of stress relaxation or in terms of volume decrease, giving the same activation parameters for these cases as is measured (Fig. 4.1) and thus showing that they are aspects of the same physical process. It follows from the theory that the temperature dependency is according to the WLF equation instead of the usual Arrhenius' equation.

The same deformation kinetics theory of course has to be able to explain Arrhenius' equation, glass transition and the WLF equation and rubbery behaviour above glass transition (Van Der Put 1991) (see Fig. 4.2 and 4.3).

The derivation of the WLF equation explains temperature shifts according to this equation by entropy changes and neglectable enthalpy changes (in accordance with rubber behaviour). Thus the very high enthalpies found being a number of times greater than the dissociation energy of the material (and thus impossible), being for instance reported for wood, is caused by applying a wrong (Arrhenius-type) operation to WLF temperature dependency. This is one of the shortcomings (giving wrong predictions and making explanations impossible) of not demanding physical reality of models.

As with every science, deformation kinetics provides exact and approximate descriptions of the behaviour and the theory is mostly applied in a too simple phenomenological form for a certain goal and

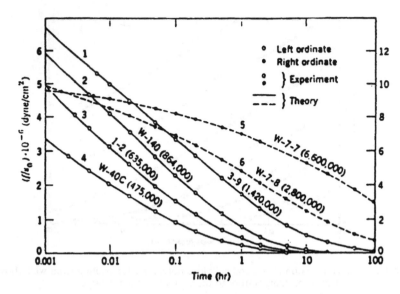

Fig. 4.3 Stress relaxation of rubbery polymers above glass transition.

it and is often rejected based on these insufficient applications. It has however to be realized that this rejection should also apply for all usually applied methods in engineering being phenomenological too and even are not able to fit a single experiment of, for instance, loading, creep, unloading and recovery. These methods thus cannot be used outside the measuring range and conditions and cannot be used for extrapolations or predictions.

For a real physical meaning the relation of deformation kinetics solutions with physical or chemical reality always has to be checked and the fit of the behaviour (for any loading history) of a specimen must be precise with a correlation close to one, only determined by the measuring precision, because the behaviour is deterministic for a given structure by the last of great numbers by the many molecules.

The activation parameters must also show this reality. For instance there must be a coincidence of activation enthalpy with the sublimation energy and lifetime and creep parameters must agree with the Debye frequency and lattice distance, etc. The theory of Van Der Put (1989) for instance shows the principal explanation by the main processes of all aspects of time-dependent behaviour of wood, showing the same activation parameters in all circumstances being the same as those of other polymers with the same type of bonding, showing the reality of these processes. The end goal is to know the nature of all imperfections in wood (all processes) making the prediction of behaviour under stress possible.

Structural materials are regarded as linearly elastic at sufficiently low stresses, temperatures and short loading times. Although the internal spring is Hookean, the measured apparent modulus of elasticity depends on viscous contributions and thus depends logarithmically on the rate of loading as explained by deformation kinetics (Van Der Put, 1989).

These viscous effects, for example, at loading to a stress level of a relaxation test have to be accounted for to be able to describe the response. Quick loading shows a higher modulus and a small quick unloading because there is an influence of a process with a small relaxation time that cannot be loaded at lower speeds. Further this quick loading causes a shorter delay time of the second process (see Fig. 4.4), as explained by deformation kinetics, causing a shift of the relaxation lines along the time-scale to shorter times while the long delay time of the third process will shift to longer times by the lower initial loading of those sites. Thus creep and relaxation are influenced by and have to be analysed for the initial conditions of these processes.

For higher stresses, temperatures and longer loading times the behaviour of structural materials is often described as linearly viscoelastic, being a severe limitation in possibilities of describing time-dependent behaviour because the stress dependency of the apparent relaxation time is one of the most striking properties of this behaviour.

At longer times or at still higher values of stress and temperature the behaviour is clearly non-linear and the future response after long periods of low stresses and temperatures can be derived from the behaviour at high stresses and temperatures because the processes acting are the same. Often the linear approach for longer times is maintained by a spectrum of relaxation times. In the literature an example

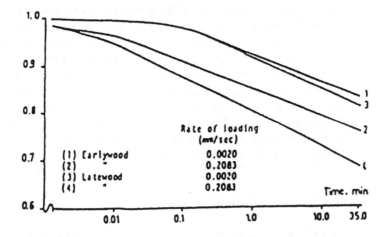

Fig. 4.4 Fractional stress relaxation s/s_0 of high loaded micro-specimens.

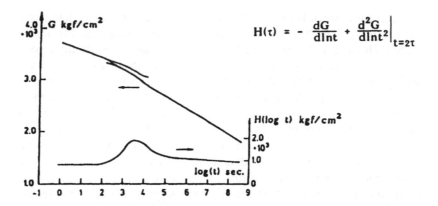

$$H(\tau) = - \frac{dG}{d\ln t} + \frac{d^2G}{d\ln t^2}\Big|_{t=2\tau}$$

Fig. 4.5 Master relaxation curve, following from horizontal shifts of relaxation lines of different temperatures and relaxation spectrum H.

is given of 40 Kelvin-elements to describe the first stage of stress relaxation while the same can be described by one non-linear element. In fact it is necessary that one real mechanism (with one activation enthalpy) determine the 'spectrum' because the spectrum as a whole shifts along the time axis at other temperatures which is impossible when different relaxation times exist.

A spectrum can be constructed using the properties of this non-linear mechanism (see Fig. 4.5) showing that shifts are different for both plateaux separated by the peak (because of the two different enthalpies of the two processes) and that the spectrum should be more dimensional (for instance a spectrum is different for every stress level) and that the use of a spectrum is a superflueds mathematical operation.

The non-linearity of the behaviour at longer times, even at low stress levels, is accepted by everyone by using power laws of stress and the therefrom following power law of time for creep and damage processes. The possibility, as now is developed for wood, of a general mathematical description of parts of the behaviour by series of non-linear functions requires almost too much parameters to be able to measure and of course it is an empty success to be able to describe phenomena with many meaningless parameters. Thus the only possibility for a complete description of time-dependent behaviour is to retain molecular reality and to use the theory of deformation kinetics.

4.2 Theory of molecular deformation kinetics

Thinking of a solid as a giant molecule highlights the fact that many types of structures and various types of bonding make up the typical solid. The plastic or viscous flow of any material is, by the nature of the process, solely the consequence of breaking bonds and establishing new bonds in a shifted position. The macroscopically observed shape change is the sum of the individual atomic events. Because of this fundamental and general fact, the process of plastic flow has to be considered as a chemical reaction in which the composition remains constant but the bond structure of the molecule changes. Thus plastic deformation is identical to the isomerization of a giant molecule – the specimen. This means that the rate of the plastic deformation or of the damage increase (when broken bonds do not re-form) is determined by the rate of this reaction and has to be described by the reaction equation.

In Van Der Put (1989) it is shown that a consistent general theory can be derived solely based on the reaction equations of the bond-breaking and re-formation processes at the deformation sites (i.e. spaces that the molecules may move into) due to the local stresses in the elastic material around these sites. The theory does not contain other suppositions and by substitution of the number and dimensions of the flow units into the concentration term and work term of the rate equation, the expressions for strain rate, creep, hardening and delay time, rubber behaviour, etc. and of the structural change processes such as fracture, fatigue, glass-transition, etc., are directly derived.

The derivation in Van Der Put (1989) is based on a Fourier expansion of the potential energy curve. Each sinus row of the expansion acts as one symmetrical energy barrier and the reaction is split in to different chemical steps (in accordance with Hess's law). Thermodynamics shows that for this type of barrier (with no change of specific heat) the activation energy is linearly dependent on temperature, moisture content and stress (while the stress independent parts of the enthalpy and entropy terms are constant with respect to temperature) providing simple equations (non-linearity of the activation parameters indicate the existence of more processes). The derivation implies the proof of the generalized flow theory, and shows that the hypotheses, on which this theory was based, are consequences of the series expansion. The activation energy and volume provide information about the type of bonds that are involved and the concentration of the flow units and these two parameters determine the mechanical behaviour in all circumstances.

The kinetic strain rate equations based on Bolzmann statistics provide the constitutive equations being implicit functions of the external stress, stress rate, strain, strain rate and of the internal stiffnesses at the sites and satisfy the general principles of physics, mechanics and thermodynamics at the sites. Outside the sites the material is elastic.

The theory is able to explain the different power models of the stress (as used in fracture mechanics and the Forintek model) and of the time (as in the Andrade and Clouser creep equations), giving the physical meaning of the exponents and constants and the change of these 'constants' in different circumstances. The theory however provides a general description of the governing phenomena not only applicable for one type of process within a limited time, loading and temperature range as where-fore these models apply.

The solutions of the deformation kinetics equations:

$$\frac{d\varepsilon}{dt} = \dot{\varepsilon} = \frac{\dot{\sigma}_i}{K_i} + (A_i + B_i\varepsilon_i)\ \sinh\ (\sigma_i\varphi_i(1 - C_i\varepsilon_i))$$

where B_i is the part of the mean stress on sites ε_i, $_i$ is the plastic strain, and K_i, is the equivalent local modulus of elasticity and A_i, B_i, C_i and φ_i are known functions of the activation parameters and temperature, are given in Van Der Put (1989) for transient processes at different loading histories, explaining the phenomenological laws, as mentioned below, for creep and fracture.

For wood it is sufficient to put $C_i = 0$ in the equation if an ultimate plastic strain condition is used and further there are only processes with A_i (and $B_i = 0$) or with B_i (and $A_i = 0$) and there are never more than two dissipative processes working (or better noticeable) at the same time.

A striking point is that the theory explains the quasi-'irrecoverable' flow as the stress dependency of the apparent relaxation time due to the non-linear behaviour providing a mobile behaviour at high stresses and a very immobile behaviour for the low internal stresses after unloading. Real irrecoverable behaviour may occur due to structural changes ($A_i = 0$; $B_i \neq 0$).

Further, a derivation of the mechano-sorptive effect in wood is possible (Van Der Put, 1989) and the behaviour at moisture cycling is explained (for example, Fig. 4.6).

The derivation throws a new light on this flow mechanism, being a

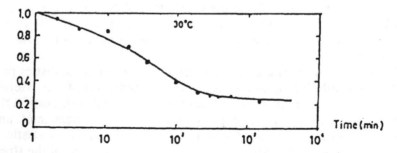

Fig. 4.6 Fractional stress relaxation at moisture content change.

separated sorption effect by, at the same time, swelling and shrinking of adjacent layers and is not an interaction effect of creep and moisture change or an interaction of loading with the overall shrinkage.

4.3 Conclusions from the application of the theory

Application of the theory leads to the following conclusions. For wood and many other materials there is at the start a dominating mechanism with a constant value of $\varphi\sigma = n$ (see model equation) independent of the temperature and initial stress s (for creep). This applies, at constant moisture content, for side-bond mechanisms and segmental movements and also for short segment movements of rubbers in the glassy state. According to the theory and measurements this viscous deformation is recoverable. Probably $\varphi\sigma$ is also constant for slip-line formation (as for metals), local buckling, and crack initiation and propagation.

Coupled to this side-bond mechanism (for wood and crystalline materials) is a structural change mechanism with a lower value of $\varphi\sigma$ and a long delay time (by flow unit increase) that occurs after some critical viscoelastic strain (about 0.4% for wood) of the first mechanism and possibly this first mechanism causes the loading, and creates the room, being the flow units, for this second mechanism (see Fig. 2.14).

The additional creep of this second mechanism is irreversible (as follows theoretically from the structural change process by the increasing flow unit density).

The dependency on the moisture content ω, following from strain rate tests, showed that $1/n$ is proportional to ω (or the density of flow units is proportional to ω). The same followed (of course) from creep tests. Besides these dominating mechanisms, that are related to the cellulose and hemicellulose, there is a small (neglectable) mechanism in the lignin with a low value of $s\varphi$ and a short relaxation time (being quasi-Newtonian) that is only noticeable at very high loading rates. For dense species lignin may have a pronounced influence (see below).

Of minor importance and mostly neglectable is a primary bond-breaking process with n $\neq$ 63 and an activation energy of about 50 kcal/mol (this process dominates in a controlled crack growth test). With the property of constant $s\omega$ (or constant n), it is possible to explain experimental laws such as, for instance:

- the linear dependence of the stiffness on the logarithmic value of the strain rate in a constant strain rate test;
- the logarithmic time law for creep and relaxation, the necessary breakdown of the law for longer times and the bend off of the creep line at long times (delay time of the structural change process);

- the shift factor along the log-time axis due to stress, moisture and temperature, for strength (not for temperature), creep and relaxation and the influence on this factor of a transition to a second mechanism;
- the yield drop in a constant strain rate test and the related delay time of the creep or relaxation test ;
- the logarithmic time to failure law in the constant loading rate test;
- the different power models (of the stress and of the time) such as the Forintek model and the Andrade and Clouser creep equations;
- the constant of the WLF- equation for glass–leather transition;
- the heights of the relaxation spectre, etc.

The exponent n of the experimental power law equation for the creep rate and the exponent n of the Forintek damage model is equal to the work parameter n of the activation energy. Further 1/n is the slope of the normalized logarithmic creep and relaxation laws and of the logarithmic time-to-failure law of the creep strength or long duration strength. The value of n is n = 34 in the Forintek model. As the slope of the logarithmic creep-to-failure law n = 38 is found, if the line is scaled to the $\neq$ 1 sec. short-term strength, but n = 34 when scaled to the 5 min. strength, according to the theory. The value of n following from the universal form of the WLF- equation as applied for the glass transition of lignin is: n = 2.3 × 17.44 = 40, indicating that scaling must be done to a very short duration strength (and probably to the dry strength) to obtain one value, showing that n essentially is a constant structure (probably unaffected by moisture content and temperature as at transition) for secondary bond processes.

The viscous flow strain at loading to the creep or relaxation level, is approximately proportional to the square of the strain $\varepsilon_v \cong c\varepsilon^2$ for holocellulose (wood structure without lignin) because of the parabolic loading curve. For wood at low stresses ε_v can roughly be assumed to be about linearly dependent on and is quasi-linear by the influence of the lignin wherefore $B\varepsilon_v = 0$ or A is constant (by the high initial flow unit density and no structural change).

For dense species with a high lignin content, an additional flow unit multiplication mechanism dominates at the start with a stress independent relaxation time and thus not related to holocellulose. It can be deduced that for this mechanism $\varphi\varepsilon_v$ or $\varphi\varepsilon^2$ is constant (because of the constant relaxation time). So the density of the flow units is proportional to the plastic strain. The constancy of $\varphi\varepsilon_v$ applies only for constant temperature and moisture content. Possibly $1/\varphi\varepsilon^2$ is linearly dependent on the moisture content ω and absolute temperature T (this has still to be investigated).

For clear wood in compression there is no indication of hardening and yield drop, showing the influence of an amorphous polymer (lignin). For wood in tension there is a high yield drop, showing the influence of a crystalline material (cellulose) dominated by a low initial flow unit density. There is also no indication of hardening for tension, showing that the change of one model parameter dominates in this structural change process as also follows from the high value of $\sigma\varphi$.

For wood the logarithmic law of the creep-to-failure strength is one line for different wood species, moisture contents (if not too high at high temperatures because of a transition), stress states (bending, shear, compression, etc.) and types of loading, indicating that $n = \varphi s$ has to be constant, independent of moisture content and also that the activation enthalpy and entropy are independent on the moisture content. The activation volume is however linearly dependent on the moisture content (and thus the inverse of the strength has the same dependence). The experimental creep-to-failure tests at different temperatures and moisture contents could be explained as well as the straight line of the strength on log-time scale for dry wood as the curved line for saturated wood, showing an enthalpy of about 36 kcal/mol above a transition temperature of - 8°C and about 30 kcal/mol below this transition temperature (determined for saturated wood because dry wood does not show this transition). Also for thermal degradation alone 36 kcal/mol was found.

The value of n is probably the result of two main processes. Values of $n \neq 62$ (used in the power law of stress) are given in literature for controlled crack growth tests, to $n \neq 65$ in constant strain rate tests, and $n \neq 30$ in creep-to-failure tests as apparent value of the first overall process. In other experiments also values of $n = 25$ to 39 are mentioned (because n in the power law of stress depends on the test). n partly also depends on the chosen scaling strength. However analysing the creep values at not too high stress, the existence of two parallel barriers was clearly demonstrated. The quick process had a high internal stress (forward activation only) and an activation energy of approximate 50 kcal/mol. The slower process was approximately symmetrical and had an activation energy of about 21 kcal/mol. The quick process, that was determining in the first stage of the loading may probably be associated with the first determining crack propagation process with $n \neq 62$ and the second process may be associated with the slower process with $n \neq 30$. The activation energy of this process is comparable with other values mentioned in the literature where from creep tests at different temperatures for bending (22 kcal/mol to 24 kcal/mol, depending on the temperature range) have been found. From normal-to-grain relaxation tests 23 kcal/mol was reported for wet beechwood. This energy is often regarded as the energy of cooperative hydrogen bond breaking. The activation energy of 50 kcal/mol is probably high enough for C-O bond or C-C bond rupture. As mentioned before a third process may start coupled to the above-mentioned slower process in the creep test at not too high stress levels. This third process has an increasing and low initial concentration of flow units causing a delay time (depending on the stress level as predicted by theory), indicating a structural change process. The activation energy of this process is 46 kcal/mol as follows from creep tests at different temperatures. The 'relaxation time' of the other process (with a high, constant flow unit density) cannot be measured accurately because of the start of the third process (even if this starts after a long time at a low stress level) and more data from different loading histories are necessary.

A rheological and strength model for wood has to contain all the above mentioned processes acting in parallel. Some processes do not follow the time–temperature or time–stress equivalence showing that the commonly applied empirical methods cannot be used. Much more work has to be done for a complete picture of this revelation of nature given by deformation kinetics.

4.4 References

Van Der Put, T.A.C.M. (1989). Deformation and damage processes in wood, Ph.D. thesis, Delft University Press, Delft, the Netherlands.
Van Der Put, T.A.C.M. (1991). Explanation of annealing of amorphous solids. Report 25.4-91-05/C/HA-51 Stevinlab. TU-Delft.

5

Present knowledge of mechano-sorptive creep of wood

D.G. Hunt

5.1 Introduction

The deflection of wood structures in service is practically important, because with the dimensions used, failure usually occurs first by unacceptably large deflections, and only at a much later stage by actual rupture.

Deflection is partly elastic and partly time-dependent creep. Besides the importance of time as a variable, it is known that changing humidity, either increasing or decreasing, causes wood to creep more than at constant humidity, the strain depending only on the amount of moisture change, not on the test time or the rate of change (Armstrong and Kingston 1962). This effect is known as mechano-sorptive creep. Altogether it has been established that the following variables are involved in wood creep:

- Wood characteristics (density, microfibril angle, elastic modulus, shrinkage rate, etc.)
- Stress
- Stress history
- Time
- Moisture content
- Moisture-content change
- Moisture-content history
- Temperature
- Temperature history (possibly).

Because of the large number of variables, and the magnitude of any experiment that could include all of them, any quantitative evaluation requires a model. A model to give quantitative results could be either one or more purely mathematical constitutive equations or a combination or a structural description with its corresponding equations.

5.2 Testing methods

5.2.1 Types of test

Measurement of 'time-dependent behaviour' is a large field and could cover:

(a) all aspects of time-dependent stress–strain behaviour, i.e. constant stress (creep), constant strain (stress relaxation) or regimes in between these;
(b) all time scales from quick high-frequency dynamic tests to long-term (quasi-static) tests;
(c) all three orthotropic directions (axial, radial, tangential or combinations);
(d) all loading modes, i.e. bending, tension, compression, shear or combinations;
(e) all sizes and qualities (clear wood, large pieces with knots and other defects, etc.);

For simplicity in the past, most measurements have been quasi-static creep tests in the axial direction under bending loads on small, clear wood test pieces. Occasional experiments have also been done in tension or compression in order to help to understand the bending results.

5.2.2 Special aspects of mechano-sorptive, compared with ordinary, creep measurements

Most of the problems with measurements of mechano-sorptive creep result from the fact that what is being measured is not a phenomenon in itself, but involves differences between phenomena. Moreover it is not a simple difference between two measurements, but is a second difference (i.e. a difference between differences). For example in the case of tension, the total creep ε_c is the difference between the measured strain ε_m of the loaded piece and the shrinking and swelling strains of a 'perfectly' matched zero-load-control piece ε_{zic} so that $\varepsilon_c = \varepsilon_m - \varepsilon_{zic}$.

However, the mechano-sorptive creep ε_{ms} is then the difference between this value and the calculated strain of a piece that was only subjected to 'normal' time- and moisture -dependent creep, ε_n, but not mechano-sorptive creep:

$$\varepsilon_{ms} = (\varepsilon_m - \varepsilon_{zic}) - \varepsilon_n.$$

In some cases this difference equation involves relatively small differences between relatively large strains, which means that standards of accuracy must be high. In the case of bending, the first difference ε_c will be the difference between the surface strain on the tension and compression faces, with the assumption that ε_{zic} is the same on each face.

As a rough rule, it may be considered that the level of accuracy that should be aimed at for research into mechano-sorptive creep should be an order of magnitude better than for research into ordinary creep.

The problem of the accuracy of second differences leads to the second main problem, namely the variability of the material. If differences are to be calculated between two measurements, then are those differences either (a) between separate and consecutive measurements on the same piece; or (b) between simultaneous measurements on two very well-matched separate pieces? In the case of (a), where the two measurements are not done at the same time, the environment must be very closely controlled in order to get a good comparison. In the case of (b), where two pieces are used, these must be well matched. However, it is well known that in a material such as wood, matching is very difficult.

Two methods of matching are 'end matching', and sorting according to elastic modulus and density; but neither method is perfect. It is also important to ensure that, if consecutive measurements are to be made on the same piece, the first measurement does not affect the properties of the piece in any permanent way.

It may be seen that great care is necessary in designing an experiment: in matching samples, in controlling the environment, and in making preliminary calibrations of test pieces and unloaded control pieces.

An example of a common problem with measurements in bending with moisture cycling is shown in Fig. 5.1. On the strain–time plot there

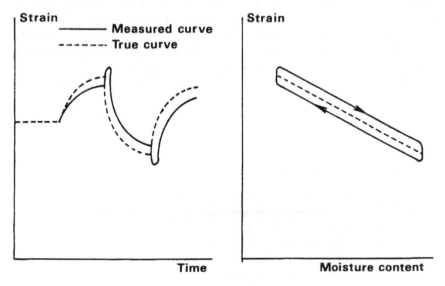

Fig. 5.1 Apparent anomalies in creep measurements during moisture cycling.

are small spikes; on the strain–moisture plot the loop shows sudden changes of strain at each end of the cycle. These can result from friction at the supports caused by the axial expansion or shrinkage of the test piece: when the piece is expanding the friction tends to increase the deflection, when it is shrinking it decreases the deflection. At each end of the cycle the movement direction changes, resulting in an apparent anomalous change in the measured deflection.

A second example of apparently anomalous behaviour also appears in Fig. 5.1 and depends on the assumptions underlying the differences discussed above. During a bending test it is usually assumed that all of the deflection is associated with elastic or creep strains, since it is expected that ε_{zic}, the zero-load swelling and shrinking, is the same on the two faces.

However, as can be seen in Fig. 5.3, this is an incorrect assumption. Failure to correct for this anomaly results in the well-known creep-recovery cycling effect during humidity cycling, shown as the dotted lines in Fig. 5.1.

A number of other corrections are also needed for good accuracy. Some of these are discussed in Hunt and Shelton (1987).

5.3   Characteristics of mechano-sorptive creep and requirements of a model

Lists of requirements of models of mechano-sorptive creep were given by Schniewind (1966) and by Grossman (1976). The following list has been revised to include the results of subsequent research. It has also been widened to include some observations on dimensional changes caused by 'stress-free' swelling and shrinkage and also certain observations on the sorption and desorption of moisture.

It was felt that these are relevant in view of the importance of mechano-sorptive effects. Except where stated, the list below refers to axial effects only, since this is usually the direction of practical significance.

5.3.1   Mechano-sorptive creep

(a) During all moisture desorptions, the creep increases faster than it would at constant moisture content, regardless of the rate of desorption (Armstrong and Kingston, (1960) and Fig. 5.2).
(b) During any first sorption at a given moisture level after loading, the creep increases faster than it would at constant moisture content, regardless of the rate of sorption (*ibid.* and Fig. 5.2).
(c) During any subsequent moisture sorptions at the same moisture

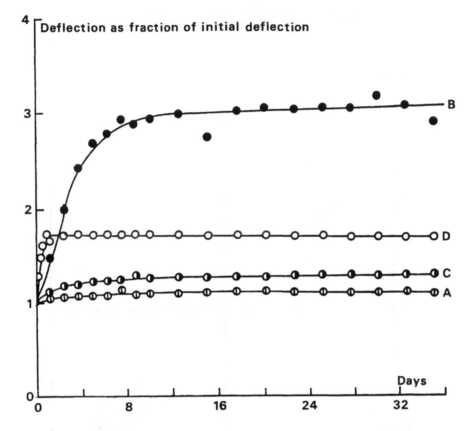

Fig. 5.2 Deflection ratio–time curves from beams at various moisture conditions.
A : Green wood kept green
B : Green wood drying to 12° moisture content
C : Wood kept at 12° moisture content
D : Wood initially at 12° moisture content allowed to absorb moisture
(Taken from Armstong and Kingston, 1962.)

level, the creep deflection usually decreases, although it may remain constant or even slightly increase (Fig. 5.2).

This apparent recovery of bending deflection during sorption has been ascribed to differential dimensional changes on the tension and compression faces (Hunt and Shelton 1988). This conclusion was supported quantitatively by measurements such as those of Fig. 5.3.

(d) As a general rule, the effect of moisture content change on the total creep deflection is much greater than the effect of time (Schniewind, (1966) and Fig. 5.4).

(e) During moisture cycling there is an 'exhaustion' process, the creep increase per cycle slowly decreasing and tending towards a limiting

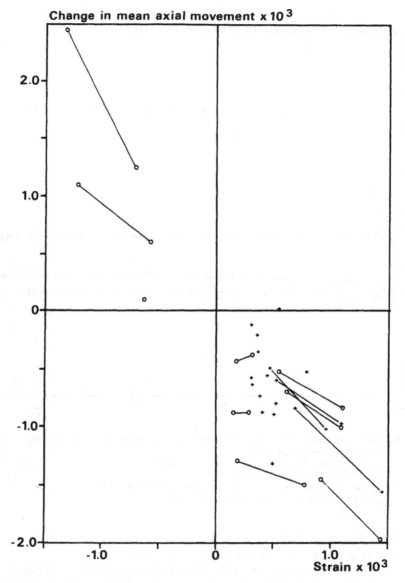

Fig. 5.3 The effect of strain level on the mean slope of strain–moisture loops.Tension and compression of bending + (tensile surface values). Lines join points obtained from the same test piece.

value beyond which it will not go (Hearmon and Paton, (1964) and Figs. 5.5 and 5.6). At this limit, the effects of 5.3.1(a) and (c) still apply: i.e. the creep increases during desorption but decreases again during sorption (Hunt and Shelton, 1988)

(f) Mechano-sorptive creep is linear at low stresses (about 10% of the

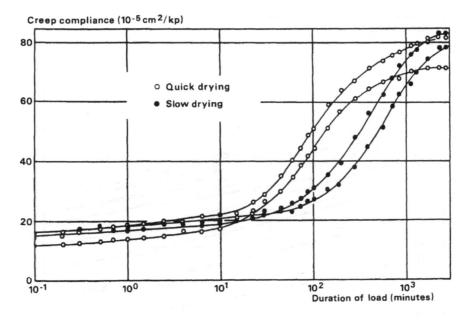

Fig. 5.4 Creep compliance for quick or slow drying. (Taken from Schniewind. (1966)).

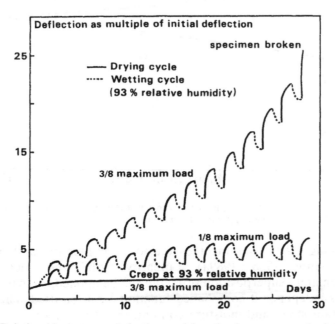

Fig. 5.5 Relationship between deflection and length of exposure cycle (taken from Hearmon and Paton (1964)).

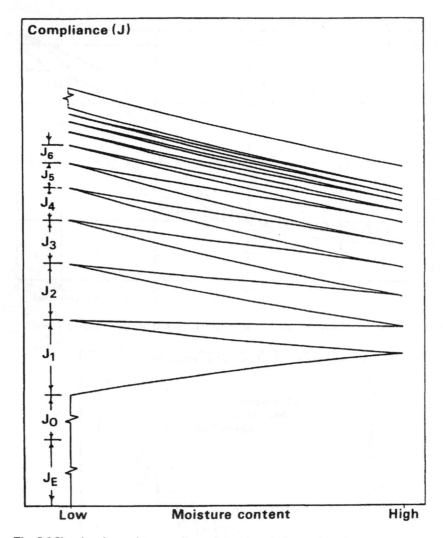

Fig. 5.6 Showing decreasing compliance increment during each moisture content cycle, with gradual approach to an apparent creep limit.

breaking load). This includes tension, compression and bending (e.g. Hunt, 1986, 1989a). Linearity with stress means that Boltzmann's superposition principle can be applied, so that a single stress-independent compliance function J can be used. Thus, with constant stress and moisture content, the strain can be calculated as:

$$\varepsilon(t) = \sigma J(u,t)$$

With changing stress but constant moisture it is:

$$\varepsilon(t) = \int_0^t J(u, t - \tau) \frac{d\sigma(\tau)}{d\tau} . d\tau .$$

With constant stress but changing moisture it is:

$$\varepsilon(t) = \sigma \int_0^t J(u, du, t - \tau) \frac{du}{d\tau} . d\tau .$$

With both stress and moisture changing it is:

$$\varepsilon(t) = \int_0^t \left[\int_0^t J(u, du, t - \tau) \frac{du}{d\tau} d\tau \right] \frac{d\sigma}{d\tau} . d\tau .$$

Stress linearity allows sign changes of stress including tension, compression and recovery to be incorporated. It also makes the calculations relatively simple for automatic computation since the strain for each new stress increment can be added to those already present.

Mechano-sorptive creep is not symmetrical in tension and compression, although the final magnitudes appear to be similar. At higher stresses it is non-linear (Hoffmeyer and Davidson, (1989) and Fig. 5.7). The strain is increasing rapidly and irrecoverably in compression zones due to cell-wall buckling damage.

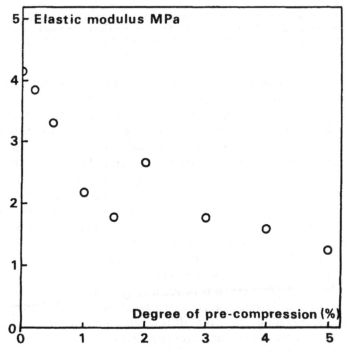

Fig. 5.7 Effect of axial pre-compression on elastic modulus (taken from Hoffmeyer and Davidson, (1989)).

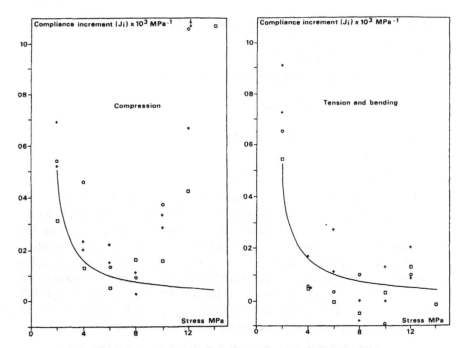

Fig. 5.8 Mechano-sorptive creep increments during successive humidity cycles when stress was increased by 2 MPa per cycle. Solid curve shows exponential approach to limit. Symbol represent different test pieces, (taken from Hunt, (1989a)).

The extent of cell-wall damage is especially sensitive to stress level (Fig. 5.8) and to high moisture contents.

(g) On removal of a load, the wood 'recovers' in a similar manner to that for normal creep, approximately according to the usual expectations of a material that is linear with stress. The elastic modulus on unloading is not measurably different from that on loading, unless the compression loading is high enough to cause compression damage. In this latter case, the elastic modulus is decreased and the zero-load dimensional change coefficient is increased (Hoffmeyer and Davidson, 1989). Recovery is speeded up by moisture changes or cycling, especially during wetting (i.e. there is a 'memory' of the original dimensions) (Ranta-Maunus (1975) and Fig. 5.9; Arima and Grossman (1978)).

(h) Mechano-sorptive creep is never a single phenomenon, but always involves a difference of at least two dimensional changes. In the case of bending it is the difference between the dimensional changes of the tensile and compressive faces; in the case of direct tension or compression it is the difference between the measured dimensional changes and those estimated to take place for the

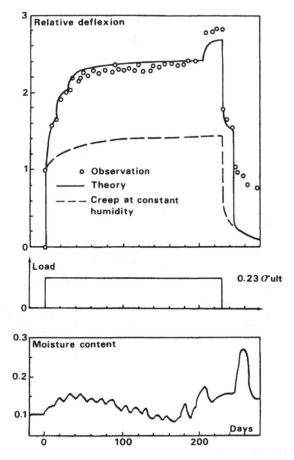

Fig. 5.9 Creep and recovery of spruce veneer in the grain direction (taken from Ranta-Maunus, (1975)).

same moisture change under zero load. If the mechano-sorptive component of creep is to be distinguished from normal creep, then a further difference calculation is needed.

(i) The mechano-sorptive effect is strongly reduced by chemical treatments such as acetylation and formaldehyde cross-linking (Norimoto *et al*, (1987) and Fig. 5.10).

This supports the theory that mechano-sorptive creep is associated with the making or breaking of hydrogen bonds under a stress bias.

On the other hand, the mechano-sorptive effect has not been observed in other hygroscopic polymers such as nylon (Hunt and Darlington (1980) and Fig. 5.11).

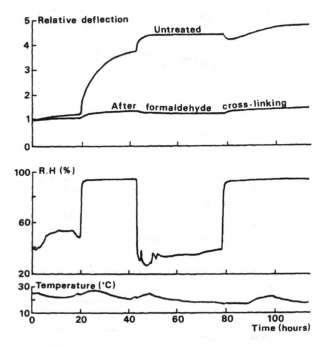

Fig. 5.10 Taken from Norimoto *et al.* (1987).

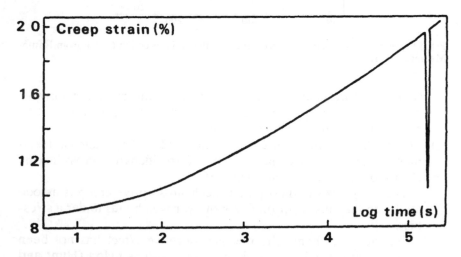

Fig. 5.11 Creep of 50 μm thick test pieces at 61% and a stress of 8 MPa. At $\log_{10} t = 5.24$, the relative humidity was reduced to 9% for 4h, and was then returned to 61%, (taken from Hunt and Darlington (1980)).

This suggests that mechano-sorptive creep in wood depends on the anatomical structure of wood in addition to the effect of stress bias on hydrogen bonds.

Also, the breaking and remaking of hydrogen bonds resulting from a constant moisture gradient through a piece of wood does not result in mechano-sorptive creep (Armstrong 1972).

(j) Mechano-sorptive creep, although mainly applying to moisture changes below the fibre saturation point, has occasionally been observed above this moisture level, mainly in species that are prone to collapse during drying, which causes an abnormal shrinkage above fibre saturation, thereby emphasizing the link between dimensional changes and mechano-sorptive creep (Armstrong 1972).

(k) The mechano-sorptive effect is also observed on loading perpendicular to the grain (Ugolev 1976), both in creep and in stress relaxation (Perkitny and Kingston 1972).

(l) The mechano-sorptive effect has been found in all species studied, any quantitative differences being probably linked to structural differences in the wood samples (e.g. Ranta-Maunus 1975).

It is also found in wood-based materials such as particle boards (e.g. Gressel 1986; Dinwoodie et al, 1990) and fibre boards (e.g. Martensson 1988).

(m) Mechano-sorptive creep susceptibility has been found to correlate positively with elastic compliance, with microfibril angle and with dimensional change rates.

This last item corroborates the findings given in (j) above.

5.3.2 Zero-load dimensional changes

(a) For any given moisture content the dimensions are larger during sorption than during desorption, thus giving a typical hysteresis curve when dimensions are plotted against moisture content (Kollmann and Cote (1960) and Fig. 5.12).

When dimensions are plotted against equilibrium relative humidity, the hysteresis is greatly reduced (Hunt (1990) and Fig. 5.13).

(b) Dimensional change rates correlate positively with microfibril angle, with elastic compliance and with mechano-sorptive creep susceptibility (Meylan 1972; Hunt 1986; Hunt and Shelton 1987 and Figs. 5.14 and 5.15).

(c) In limiting creep states (tension or compression) the dimensional change rate is inversely proportional to the relative strain; being lower under tensile strains and higher under compressive strains than in the stress-free condition (Hunt and Shelton (1988) and Fig. 5.3).

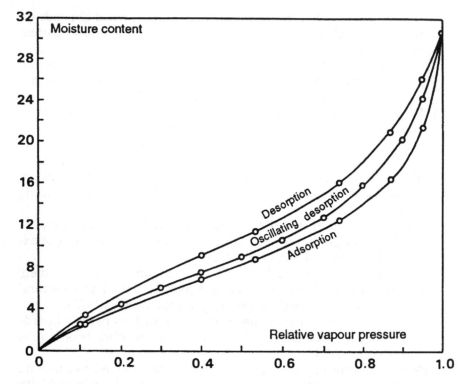

Fig. 5.12 Adsorption–desorption hysteresis curves for spruces at 25°C (taken from Kollman and Cote (1968)).

(d) With very thin pieces following step increases in the relative humidity, there is a short period of rapid expansion (or shrinkage) followed by a much longer period of constant dimensions or often of strain reversal (probably a relaxation), see Fig. 5.16.

The same effect is observed under a direct load, the difference between the two cases constituting a mechano-sorptive effect. Similar effects are observed following step decreases in relative humidity (Hunt, 1989b).

(e) Dimensional changes do not correlate well with moisture content changes. For instance during the 'relaxation' period (in (e) above) the moisture content may be increasing while the dimensions are decreasing, and vice versa.

5.3.3 Moisture sorption and desorption

(a) For a given relative humidity the equilibrium moisture content is always higher during desorption than during sorption. This was explained quantitatively by Barkas (1949) in his thermodynamic

(a)

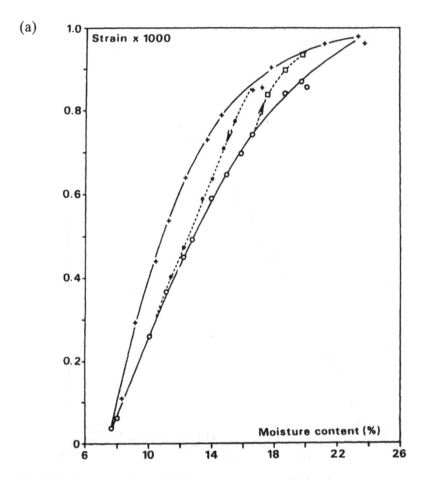

Fig. 5.13a Taken from Hunt (1990).

theory. This theory postulated that a tensile stress increased the equilibrium moisture content and a compressive stress decreased it. The theory was well supported by experimental measurements perpendicular to the grain, although parallel to the grain it predicted only small changes.

(b) Many attempts have been made to apply Fick's first and second laws of diffusion to the diffusion of water vapour in wood. Some success has been obtained with the first law, but successes with the second law have been limited. Numerous correction factors and other methods have been tried to explain the discrepancies. Some of these discrepancies have been ascribed to size effects and moisture history effects. It would be helpful if the second law could be used to predict moisture variations in wood in a known temperature and relative

(b)

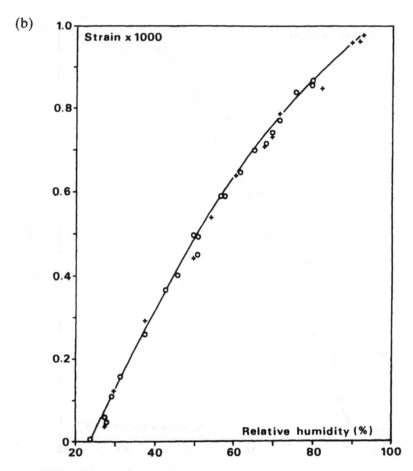

Fig. 5.13b Taken from Hunt (1990).

humidity environment, thus leading to a prediction of mechano-sorptive creep. However, in the present state of knowledge, such predictions might be inaccurate (see Walker (1977) for literature survey).

(c) During moisture sorption and desorption, the differential rates of expansion and shrinkage in different macro-regions of the material must result in the build-up of internal stresses, which must then affect the mechano-sorptive creep (Thelandersson (1990); Ranta-Maunus (1989)).

5.3.4 Dynamic effects

A number of rheological studies of wood have been made by means of dynamic measurements. Generally these are useful in studying

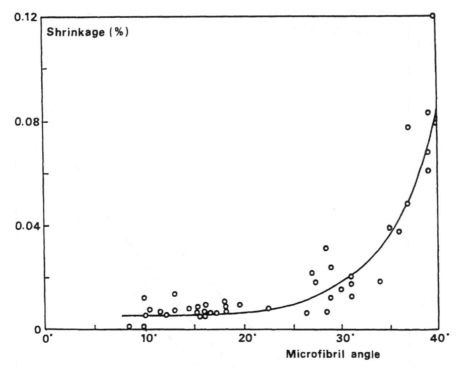

Fig. 5.14 The relationship between longitudinal shrinkage during absorption and microfibril angle at 100% moisture content (taken from Meylan (1972)).

transitions and structural effects from the point of view of wood science. In particular, transient decreases in stiffness were observed during moisture sorption (Back *et al.* 1985). However, in general the conclusions from dynamic measurements cannot be directly used in the design of wood structures nor in the prediction of their service performance.

5.4 Structural theories and constitutive equations

Very many authors have made suggestions and given partial explanations of the dimensional changes of wood under load during moisture changes. Due to the limitations of space, only ten of the more elaborate studies are included here in alphabetical order.

Bazant (1985) proposed a constitutive law based on the thermodynamics of the diffusion processes. Creep under constant conditions used the Maxwell chain model with variable coefficients; the additional creep under variable conditions was caused by stress-induced dimensional changes. Although this theory resulted in a set of equations,

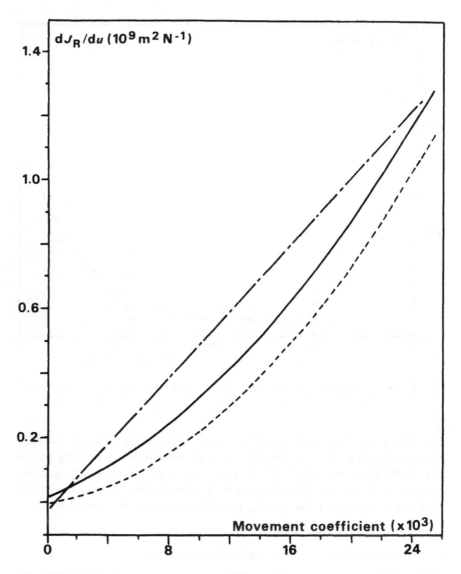

Fig. 5.15 Relation between creep susceptibility and free movement coefficient (taken from Hunt and Shelton (1980)).

there are problems with determining the various numerical parameters needed for their evaluation.

Boyd (1982) simplified the structure of wood into a basic single cell consisting of a relatively inert envelope enclosing a water-sensitive gel. Dimensional changes of the structure were explained by dimensional changes of the gel during moisture changes or load applications,

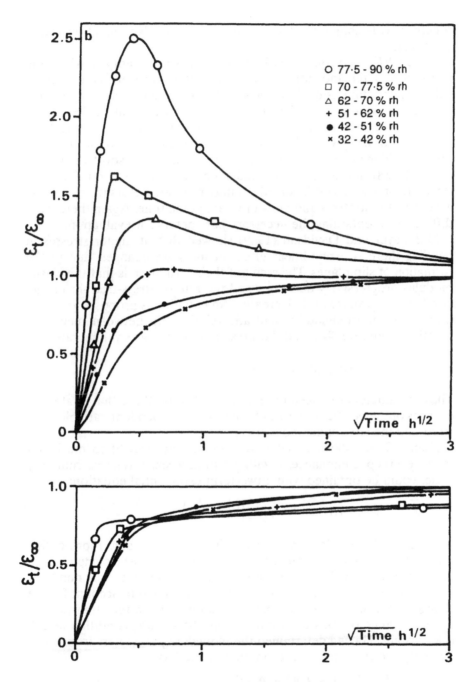

Fig. 5.16 Expansion of thin pieces during sorption steps: upper, under zero load; lower, with tensile stress of 7.5 MPa. (taken from Hunt (1989b)).

together with shape changes of the envelope. No equations were developed.

Cave (1972) and Meylan (1972) used a mixture theory in which the properties of the individual constituents of wood were combined to calculate the dimensional changes of the wood as a whole, generally without an externally applied stress. This procedure was further developed by Koponen et al. (1989) who calculated elastic and shrinkage properties of earlywood, latewood and compression wood in the axial direction. None of these authors considered mechano-sorptive creep.

Gril (1988) developed a thermodynamic theory based on the ultrastructure of wood. This involved a number of parameters which change according to the stress and moisture state at the time. Again there were difficulties in obtaining the necessary parameters for evaluation.

Hoffmeyer and Davidson (1989) showed that at higher stresses in bending there is a non-linear effect of moisture changes, due to the formation of slip planes. The result of their formation is that the elastic modulus progressively decreases and the rate of stress-free shrinkage/swelling rate progressively increases. An equation is proposed in which there is an approximately quadratic relationship between mechano-sorptive creep rate $\dot{\varepsilon}_{msg}$ and the stress-free shrinkage swelling rate $\dot{\varepsilon}_s$:

$$\dot{\varepsilon}_{msg} = \dot{\varepsilon}_s \left(1 + b\sigma |\dot{\varepsilon}_s| \right) + a\sigma |\dot{\varepsilon}_s|$$

These parameters still need to be evaluated, as do those that relate the changes of the elastic properties to the basic independent variables of stress and the environment history.

Hunt (1989a) showed that a very close empirical fit to mechano-sorptive creep compliance J with uniformly sized relative humidity cycles could be obtained with a two-term exponential equation:

$$J = J_E + J_0 + J_1 \left(1 - e^{-n/N_1} \right) + J_2 \left(1 - e^{-n/N_2} \right)$$

where J_E is the elastic compliance, J_0 is a constant value representing normal time-dependent creep, J_1 and J_2 are characteristic compliance, n is the number of cycles and N_1 and N_2 are characteristic cycle numbers. A further correction factor of $(1 + A\Delta u)$ was needed to account for the change of the shrinkage–swelling rate resulting from the strain, where u is the change of moisture content from the loading condition and A is a constant. Having determined the individual parameters from a few humidity cycles, a creep limit can be predicted as:

$$J_\infty = (J_E + J_0 + J_1 + J_2)(1 + A\Delta u)$$

In the case of non-uniform humidity cycle, it was suggested that the equation should be modified to

$$J = J_E + J_0 + J_1\left(1 - e^{-\Sigma|S_u|/u_1}\right) + J_2\left(1 - e^{-\Sigma|S_u|/u_2}\right)$$

where S_u represents any step change in moisture content. This modification has not been tested, and it also suffers from the shortcoming that it does not consider the acceleration of mechano-sorptive creep when the humidity is taken to a higher level than previously experienced since loading. This latter factor was considered by Toratti (1990).

Mukudai and Yata (1987) developed a theory based on the assumption of slippage between the S_1 and S_2 layers of the cell wall at certain moisture content levels. The effect of an applied stress was to cause slippage in the direction of the stress (i.e. like fibre slippage of a fibre-reinforced material). The model is shown in Fig. 5.17. In addition to the usual springs and dashpots it also includes swelling and shrinkage elements A_1 and A_2, a running block B running on bar D, and pinned connections F_1 and F_2. The resulting equations (too complex to be given here) generally gave qualitatively correct results using suitable parameters. It is difficult to assess the comparison with the experimental results since the latter were conducted at high stress levels at which non-linear behaviour would be expected. The published curves also gave comparative results in tension and compression that are at variance with previous experimental findings that mechano-sorptive creep in compression is equal to or greater than that in tension.

Ranta-Maunus (1989) proposed an equation that gave mechano-sorptive creep that was linear with moisture change, but had coefficients that took different values according to the type of moisture change. The simplified version (very suitable for computer simulation) was that the creep compliance J is given by:

$$J = J_E + J_N + J_{E0} \sum_{i=1}^{n} a(u_i - u_{i-1})$$

where J_E is the elastic compliance, J_N is the 'normal' creep compliance integrated over the moisture contents and times of the test, i.e.

$$J_N = \int_0^t \frac{dJ(u,\tau)}{d\tau}d\tau$$

J_{E0} is the reference elastic compliance at 0% moisture content, a is a hydroviscoelastic constant taking the values a^- or a^{++} according to whether the moisture content u_i is less than or greater than u_{i-1}, respectively. For subsequent moisture increases a^{++} is replaced by a^+. The bracketed term $(u_i - u_{i-1})$ represents step changes in moisture content. It should be noted that unless the 'hydroviscoelastic constants' are made to progressively decrease with increasing compliance, there will

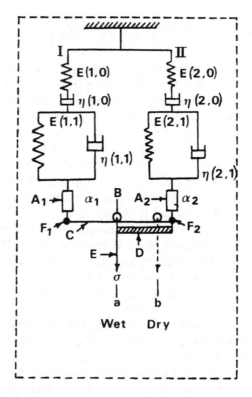

Fig. 5.17 Model representing viscoelastic behaviour of wood under moisture change.
I: Model I(S2+S3 layer)
II: Model II (I+P+S$_1$ layer)
A1,A2: Hygroscopic materials having swelling and shrinkage coefficient x$_1$, x$_2$ respectively
B: Running blocks
D: Hygroscopic materials adjusting stress bias by swelling and shrinkage
F1,F2: Pin connections
(taken from Mukudai and Yata (1986)).

be no creep limit, but a continuing increase in creep with moisture changes.

Toratti (1990) proposed a 'linearized model' incorporating some components from each of those described by Hunt and Ranta-Maunus. His equation for the mechano-sorptive creep compliance J_{ms} was:

$$J_{ms} = J_E \big[d|\Delta u| + e|du| \big] + \int (\alpha - bJ_{tot})du$$

where J_E is the elastic compliance, Δu is the moisture change to moisture contents not previously attained during the loading history, du is any moisture change, J_{tot} is the total compliance excluding shrinking/

swelling changes, and d, e, α, and b are constants. This model was compared with those of Hunt and Ranta-Maunus, but no firm conclusions were drawn.

Van Der Put (1989) proposed a thermodynamic theory to describe creep in constant or changing environmental conditions. The theory is based on the assumption of the breaking and re-forming of bonds due to sorption and desorption under a stress bias, and is supported by the necessary equations of molecular deformation kinetics. The model predicts that the mechano-sorptive effect is of minor importance in pieces of relatively large dimensions, where moisture changes are necessarily slow.

5.5 Conclusions and recommendations for future research

From the above information it is clear that much work has been done and some progress has been made on studying and quantifying mechano-sorptive creep in clear wood. However, the final goal must be to quantify mechano-sorptive creep in structural timber. Structural timber involves wood with defects, and especially with the quantity and size of defects that are the controlling factors in reliability-based design (for example the 'lower fifth percentile'). It seems reasonable to suppose that the chief defects to be considered are the same ones as mainly control fracture, namely knots and sloping grain. It may also be necessary to consider juvenile wood as a 'defect'. Very little work has been done in assessing the relative importance of these defects.

A design code must include two separate and equally important aspects. Firstly it is important to estimate the performance of a given design: this is where constitutive equations such as those given in the previous section can be useful. Secondly there must be a selection and grading procedure for a variable material such as timber. Some suggestions for this have been made for clear wood (Hunt 1986), but it is very important to extend this work to include wood with defects, i.e. timber.

One possible approach to by-passing some of the design problems could be to use the idea of a creep limit, if it exists for full-sized timber. If the creep-limit compliance can be correlated with a simple measurement such as elastic modulus, this could eliminate time, moisture content and moisture-content history from the list of independent variables, leaving only stress, stress history, timber quality and possibly temperature.

It is clear, then, that there are still many questions to be answered.

The following list includes some of the most fundamental ones:

(a) What is the relative importance of defects such as knots, sloping grain and juvenile wood?
(b) Does a creep limit really exist for timber containing defects?
(c) What is the effect of temperature on mechano-sorptive creep, and a creep limit?
(d) What is the basis for the selection and grading of timber for design against mechano-sorptive creep?

5.6 References

Arima, T. (1967). The influence of high temperature on compressive creep of wood. *J. Jap. Wood Res. Soc.* **13**, 36–40.

Arima, T. and Grossman, P. (1978). Recovery of wood after mechanosorptive deformation. *J. Inst. Wood Sci.* **8**, 47–52.

Armstrong, L. (1972). Deformation of wood in compression during moisture movement. *Wood Sci.* **5**, 81–86.

Armstrong, L. (1983). Mechano-sorptive deformations in collapsible and non-collapsible species of wood. *J. Inst. Wood Sci.* **13**, 206–211.

Armstrong, L. and Kingston, R. (1960). The effect of moisture changes on creep in wood. *Nature* **185**, 862–863.

Armstrong, L. and Kingston, R. (1962). The effect of moisture content changes on the deformation of wood under stress. *Austral. J. Appl. Sci.* **13**, 257–276.

Bach, L. (1965). Non-linear mechanical behaviour of wood in longitudinal tension. Ph.D. Thesis, Syracuse University, USA.

Back, E., Salmen, L. and Richardson, G. (1985). Means to reduce transient effects on mechanical properties of paper during moisture. *J. Pulp Paper Sci.* **11**, 1–5.

Barkas, W. (1949). *The swelling of wood under stress*, HMSO, London.

Bazant, Z. (1985). Constitutive equation of wood in variable humidity and temperature. *Wood Sci. Tech.* **19**, 159–177.

Boyd, J. (1982). An anatomical explanation for viscoelastic and mechano-sorptive creep in wood, and effects of loading rate on strength. In Baas, P. (ed.) *New Perspectives in Wood Anatomy*. Nijhoff/Junk, The Hague pp. 171–222.

Cave, I. (1972). Swelling of a fibre-reinforced composite in which the matrix is water reactive. *Wood Sci. Technol.* **6**, 157–161.

Christensen, G. (1962). The use of small specimens for studying the effect of moisture content changes on the deformation of wood under load. *Austral. J. Appl. Sci.* **13**, 242–256.

Dinwoodie, J. (1968). Failure in timber. Part I. Microscopic changes in the cell-wall structure associated with compression failure. *J. Inst. Wood Sci.* **4**, 37–53.

Dinwoodie, J., Higgins, J., Robson D. and Paxton B. (1990). Creep in chipboard: 7. Testing the efficacy of models on 7–10 years' data and evaluating optimum period of prediction. *Wood Sci. Technol.* **24**, 181–189.

Gressel, P. (1986). Vorschlag einheitlicher Prufgrunsatze zur Durchfuhrung und Bewertung von Kriechversuchen. *Holz Roh- Wefstoff* **44**, 133–138.

Gril, J. (1988). Une modélisation du comportement hygro-rhéologique du bois à partir de sa microstructure. Doctoral Thesis, University of Paris.

Grossman, P. (1976). Requirements for a model that exhibits mechano-sorptive behaviour. *Wood Sci. Technol.* **10**, 163–168.

Hearmon, R. and Paton, J. (1964). Moisture content changes and creep of wood. *For. Prod. J.* **14**, 357–359.

Hoffmeyer, P. and Davidson, R. (1989). Mechano-sorptive creep mechanism of wood in compression and bending. *Wood Sci. Technol.* **23**, 215–227

Hunt, D. (1986). The mechano-sorptive creep susceptibility of two softwoods and its relation to some other materials properties. *J. Mater. Sci.* **21**, 2088–2096.

Hunt, D. (1989a). Linearity and non-linearity in mechano-sorptive creep of softwood in compression and bending. *Wood Sci. Technol.* **23**, 323–333.

Hunt, D. (1989b). Two classical theories combined to explain anomalies in wood behaviour. *J. Mater. Sci. Letters* **8**, 1474–1476.

Hunt, D. (1990). Longitudinal shrinkage-moisture relations in softwood. *J. Mater. Sci.* **25**, 3671–3676.

Hunt, D. and Darlington, M. (1980). Creep of nylon during concurrent moisture changes. *Polymer* **21**, 502–508.

Hunt, D. and Shelton, C. (1987). Progress in the analysis of creep in wood during concurrent moisture changes. *J. Mater. Sci.* **22**, 313–320.

Hunt, D. and Shelton, C. (1988). Longitudinal moisture-shrinkage coefficients of softwood at the mechano-sorptive creep limit. *Wood Sci. Technol.* **22**, 199–210.

Keith, C. (1972). The mechanical behaviour of wood in longitudinal compression. *Wood Sci.* **4**, 234–244.

Kitahara, R. and Yukawa, K. (1964). The influence of the changes of temperature in creep in bending. *J. Jap. Wood Res. Soc.* **10**, 169–175.

Kollmann, F. and Cote, W. (1968). *Principles of Wood Science and Technology*. Springer-Verlag, New York.

Koponen, S., Toratti, T. and Kanerva, P. (1989). Modelling longitudinal elastic and shrinkage properties of wood. *Wood Sci. Technol.* **23**, 55–63.

Martensson, A. (1988). Tensile behaviour of hardboard under combined mechanical and moisture loading. *Wood Sci. Technol.* **22**, 129–142.

Meylan, B. (1972). The influence of microfibril angle on the longitudinal shrinkage-moisture content relationship. *Wood Sci. Technol.* **6**, 293–301.

Mukudai, J. and Yata, S. (1986). Modelling and simulation of viscoelastic behaviour of wood under moisture change. *Wood Sci. Technol.* **20**, 335–348.

Mukudai, J. and Yata, S. (1987). Further modelling and simulation of viscoelastice behaviour of wood under moisture change. *Wood Sci. Technol.* **21**, 49–63.

Mukudai, J. and Yata, S. (1988). Verification of Mukudai's mechano-sorptive model. *Wood Sci. Technol.* **22**, 43–58.

Norimoto, M., Gril, J., Minoto, K., Okamura, K., Mukudai, J. and Rowell, R. (1987). Supression of creep of wood under humidity change through chemical modification. *Mokuzai Kogyo* **42**, 504.

Perkitny, T. and Kingston, R. (1972). Review of the sufficiency of research on the swelling pressure of wood. *Wood Sci. Technol.* **6**, 215–229.

Ranta-Maunus A. (1975). The viscoelasticity of wood at varying moisture content. *Wood Sci. Technol.* **9**, 189–205.

Ranta-Maunus, A. (1989). Analysis of drying stresses in timber. *Paperi ja Puu* **71**, 1120–1122.

Schniewind, A. (1966). Uber den Einfluss von Feuchtigkeitsanderungen auf das Kriechen von Buchenholz quer zur Faser unter Berucksichtung von Temperatur und Temperaturanderungen. *Holz Roh- Werkstoff* **24**, 87–98.

Thelandersson, S. (1990). Unpublished Notes for RILEM TC112, London.

Toratti, T. (1990). A cross section creep analysis. IUFRO Timber Engineering Conference, St John, Canada.

Ugolev, B. (1976). General laws of wood deformation and rheological properties of hardwood. *Wood Sci. Technol.* **10**, 169–181.

Van der Put, T. (1989). 'Deformation and damage processes in wood'. PhD Thesis, Delft University Press, Delft, Netherlands.

Walker, I. (1977). Changes of bound water in wood. *New Zealand J. Sci.* **20**, 3–10.

6

Time-dependent slip of joints in timber engineering

P. Morlier

6.1 Introduction

All design engineers are aware of the great influence of joint stiffness on deformation of timber on composite structural members. Three types of joints may be employed: mechanical connectors, glued joints and glued mechanical connectors.

Structural glued joints are relatively rigid, while mechanical connectors provide only semi-rigid joints which allow some displacement of wood members at the joint caused by a combination of connector embedment and friction between the connected members.

During the life of the structure, joints support different types of service loads; namely, permanent loads, loads of various (long, medium, short) duration, dynamic loads, etc.

Thus it is important to know how joints react to complex service loading and the associated environment (temperature, humidity) conditions. Mechanical joints are more sensitive to these factors than solid wood members, because the loads are transferred between members over relatively small contact areas of metal pins or rocks inducing high local stress concentrations in wood members. Furthermore, pin-type joints also open the interior of wood members to environmental influences.

A combination of the above factors is responsible for the highly non-linear short-term joint slip and long-term joint creep.

6.2 Wood connector rheology

Let us examine the behaviour of a simple round nail connection. Fig. 6.1, from Whale (1988), shows that as soon as the nail has been driven in, the wood is compressed and densified (black zone), in a direction perpendicular to the grain, while a crack may open parallel to the grain.

Once the nail is embedded wood compression and crack propagation both continue: this is why mechanical joints are very sensitive to environmental variations and load duration.

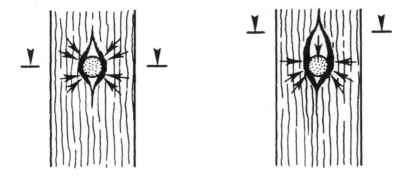

Fig. 6.1 Splitting forces around nails (after Whale (1988)).

Fig. 6.2, from Jang and Polensek (1989) shows a nailed connection between panel and solid wood. Even under a small load, the thin nail is plastically bent. At the same time, the nail pushes both on the wood and on the panel as described before, generating high local compression. These two elementary phenomena are responsible for the non-linear effects in mechanical wood joint behaviour. Fig. 6.2, also illustrates the friction between the two connected elements and it helps to visualize nail popping as nail withdrawal caused by the swelling of joint members.

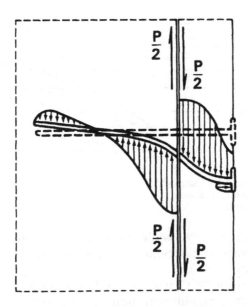

Fig. 6.2 Wood resistance to nail deformation (after Jang and Polensek (1989)).

In order to simplify the description, consider tensile joints only, ignoring moment-resistant connections here.

Obviously modelling of joints will not be easy, because it demands a complete knowledge of wood rheology, and of limit states concerning plasticity in compression, densification, crack propagation and plastic bending of connectors.

Remember also that the deformation of mechanical joints, even in the early stage of loading, tends to be non-linear because of wood damage.

6.3 Literature review

A simplified approach to predicting delayed slip in timber joints may be found in Eurocode 5, which defines a creep coefficient :

$$1 + k_{creep}(t) = \frac{\text{slip } g(t)}{\text{[elastic] slip } g(0)}$$

This coefficient is estimated for short (6 weeks), medium (6 months) and long (10 years) load duration with an instantaneous slip measured in a few hours. Three environmental classes are proposed (Fig. 6.3) for assigning design strength values and for estimating creep:

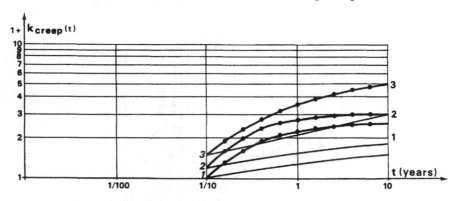

Fig 6.3 Values of k creep (in Eurocode 5).
——— bending element
—•— joints

1. The temperature is 20°C, while the relative humidity rarely exceed 65% (the average equilibrium moisture content in most softwood remains less than 12%).
2. The temperature is 20°C, while the relative humidity rarely exceed 85% (the average equilibrium moisture content in most softwood remains less than 18%).

3. Temperature and humidity conditions that lead to higher equilibrium moisture content levels than the previous service class.

The creep of wood members joined by mechanical connectors under long-term loading was studied in different countries by numerous authors (some of them are mentioned below).

Under loads not exceeding one quarter of the short-term ultimate capacity of joints, specimens rarely ruptured for duration ranging from 1 to 15 years. Some of the most important data on creep are shown in Fig. 6.4, superimposed on the Eurocode 5 recommendations:

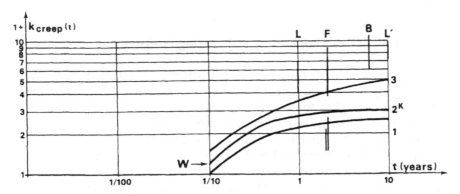

Fig. 6.4 Comparison of some experimental results with draft recommendations in Eurocode 5

       B Brock (1968)
       F Feldborg and Johansen (1986)
       K Kuipers (1977)
       L Leivo (1989)
       L' Leijten (1988)
       W Whale (1988)

Creep-rupture behaviour of selected joint types in tension has also been investigated (Palka 1989b; Foschi 1974), under moderate to high constant loads.

6.4 Tests under constant loading and constant environment

Brock (1968) gives creep results for nailed joints with different species; reporting k creep coefficient between 6 and 11 for tests of 6 years duration, he found double shear joints to be stiffer than single shear ones.

According to Mack (1963), deformations take place during the first year for nailed wood/wood joints. Utilization of joints made with green wood leads to k_{creep} up to 10. Joint creep has a large non-linear tendency, being proportional to the square root of the load. Finally, he points out the difference between pure creep (under constant load) and oligocyclic creep, which is more important in practice.

Mohler found that wooden elements are not to be nailed in a wet state, if the frame is used in a dry environment, because the resulting delayed slips are much higher than those for a structure built with dry elements.

Kuipers (1977) performed tests on different mechanical connectors during about 10 years, at 60% of the ultimate loading. The proposed k_{creep} coefficients are lower than previously, because the instantaneous slips, depending on the joint tightening, were higher than before.

6.5 Tests under varying loading and varying environment

Leicester *et al.* (1979) tested the fire resistance of different mechanical joints. They found that nailed joints are less sensitive to rising temperature, as opposed to tooth plate connectors.

Effects of humidity variations are very severe for wooden materials as observed earlier by Noren (1968). Curves obtained by Feldborg and Johansen (1968) shown in Fig. 6.5 reveal mechano-sorptive effect for joints. Recent results from Leivo (1989) are worth examining. According to Leicester *et al.* (1979) the influence of humidity variation on the delayed slip strongly depends on the species tested.

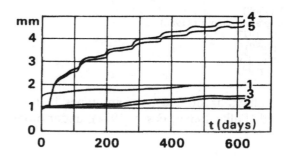

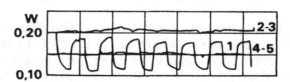

Fig. 6.5 Mechano-sorptive effect for joints (after Feldborg and Johansen (1986): slip (mm) is given versus time (days) for plywood gussets with annularly threaded nails, under constant load and constant (1, 2, 3) or varying (4, 5) environment.

The extreme scattering of values observed in Fig. 6.4 can be explained by a combination of many different factors, such as: the definition of instantaneous or elastic slip varies between authors; numerous tests were performed under uncontrolled ambient temperature and humidity conditions; it is difficult to know whether the pure deformation of the joint was measured without any contribution from the deformation of the joined elements, second order effects are present due to possible non-planar forces; sample preparation and conditioning procedures are not given, the mechanical properties for a given species are variable, the joint slip is non-linear with respect to the load, and so on.

To illustrate the last two items, consider the following experimental results: Fig. 6.6, from Whale's dissertation (1988), shows the extreme and mean creep curves for 16 simplified joints, that were supposed to be identical, at 25% of their ultimate load capacity, and Fig. 6.7, from Wilkinson (1988), was obtained on symmetrical bolted joints loaded at different percentages of the ultimate short-term load (30%, 60%, 80%), demonstrating that k creep increases with the loading ratio.

Such a complexity has induced different researchers, like Leicester *et al.* (1979) or Whale (1988) and Whale and Smith (1989), to propose a simplified procedure – the embedment test of Fig. 6.8, which allows the

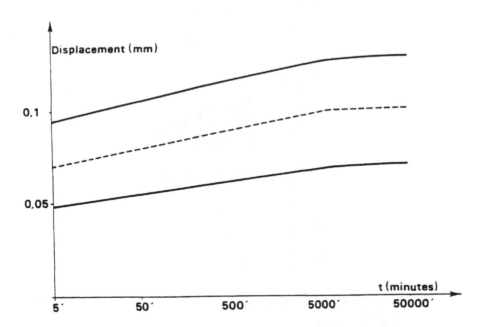

Fig. 6.6 Scattering of creep results for a simplified joint (after Whale (1988)).

------ mean value

▬▬▬ extreme values

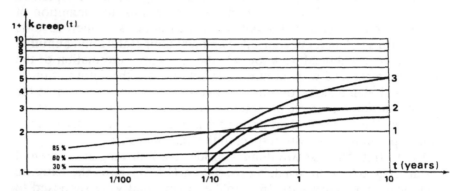

Fig. 6.7 Influence of load levels (percentage of ultimate load) on the first year creep of bolted joints (after Wilkinson (1988)).

Fig. 6.8 Embedment test (after Whale (1988))
    1 connector
    2 metallic device
    3 wooden material.

introduction of any type of mechanical loading, or any environmental variation, in order to better understand phenomena like (non-) linearity, (non-) reversibility, creep, and so on.

6.6 Creep modelling

The aim of modelling is to describe mathematically with a good accuracy the delayed joint slip in different situations: creep, cyclical loading, complex history, and so on.

Noren (1968) was the first to define k_{creep} in a classical way, corresponding to a simple rheological model.

$$1 + k_{creep}(t) = slip(t) / slip \ (t = 0.5 \ day)$$
$$= 0.385 \ (1 - e^{-t/24}) + 5.5 \times 10^{-4} t$$

where t is time measured in days.

If t = 10 years, k_{creep} = 3.25 this value can be plotted on Fig. 6.4.

Later, Jenkins *et al.* (1979) proposed a method to evaluate the slip of joints under an increasing load, when creep under different loads is known; this method is based on a pseudo-principle of superposition.

Jang's dissertation, quoted in Jang and Polensek (1989) proposes a realistic creep model for joints based on the following joint slip components:

$$G \ (t) = G \ (0) \ elastic + G \ (0) \ non\text{-}recoverable$$
$$+ G \ (t) \ recoverable + G \ (t) \ non\text{-}recoverable.$$

Two alternatives were considered. The first form is:

$$G(o) \ elastic \ = P \ / \ E_1 \ or \ B_1 P^{N_1}$$

$$G(o) \ elastic \ non\text{-}recoverable \ = BP \ or \ B_5 P^{N_4}$$

$$G(t) \ recoverable = \frac{P}{E_2} \left[1 - \exp(-\frac{E_2}{\eta_2})t \right] \ or \ B_2 \left[1 - \exp(-B_3 t) \right]$$

$$G(t) \ non\text{-}recoverable = APt^M \ or \ B_4 P^{N_2} t^{N_3}$$

where P is the load, t the time.

The second form is:

$$G(t) = \sum_{i=1}^{3} F_i{}'(t) P^i + \sum_{i=1}^{3} F_i{}''(t) P^i$$

where F' concerns the recoverable slip and F" the unrecoverable one.

This proposal shows correctly the non-linear and practically unre-coverable character of slip in mechanical connectors. For a given complex loading history, Jang and Polensek rely on the following theoretical tools: the modified principle of superposition by Nolte and Findley and strain-hardening principle by Cho and Findley.

Palka (1989a) proposed a similar approach, only slightly different in its formulation, based on earlier work by Marin.

Whale's papers (1988) moved these ideas forward, after numerous, long and complex (as far as the loading history is concerned), experimental tests referenced in Fig. 6.8, under constant environmental conditions and under loads much lower than the ultimate joint capacity. The resulting joint slip is modelled by the following equations under any loading history:

g (t) = g elastic + g (t)recoverable + g (t)non-recoverable.

If we call P ult the ultimate load,
$\qquad$ P*(t) the highest load ever reached at time (t),
$\qquad$ p = P/P ult, and
$\qquad$ p*(t) = P*(t)/P ult, then

$$\begin{cases} g \text{ elastic} = 0.12\, p^{1.3} + 0.4\, p^{*1.5} \\ \text{non linear elasticity} \end{cases}$$

$$\left\{ g(t) \text{ recoverable} = \int_0^t 0.018 \left| 1 - e^{-(0.004/0.018)\,(t-\tau)} \right| \frac{\partial p(\tau)}{\partial \tau}\, d\tau \right.$$

classical viscoelasticity

$$G(t) \text{ non recoverable} = \int_0^t J_f\left(P^*(\tau), t-\tau\right) g_1\left(\frac{p^*(\tau)}{dp^*(\tau)/d\tau}\right) \frac{\partial p^*(\tau)}{\partial \tau}\, d\tau$$

J_f is a creep function, depending on P*, which is given in Whale's dissertation together with g_1. Note that g is given in mm and t in hours.

The importance of the highest load P* ever supported by the joint during its history is evident in his formulation. Moreover, Whale suggests that in order to decrease the delayed slip of a joint, it should be subjected to a brief and relatively high load before the final loading of the structure.

Finally it may be mentioned that joint slip could be predicted from damage accumulation models also (Palka 1988a), from initial creep to creep-rupture. The contribution of Leijten (1988) to this subject is also interesting.

6.7 Conclusions

Mechanical joints in timber construction display a highly non-linear, plastic and non-linearly viscoelastic slip-load diagram.

Long-term deformation of wooden structures is governed by creep of joints, whereas short-term deformation is dominated by members in bending. For hyperstatical structures, load transfer toward bending elements occurs with time.

Humidity cycles around joints cause reduction in both the stiffness and the strength, especially for initially green jointed elements.

Most of the referenced papers were limited to simple experimental situations. Therefore, predicting the behaviour of moment-resisting joints, or the behaviour of joints under cyclic loading, and cyclic environmental conditions, must wait for future developments.

References

Atherton, G.H., Rowe, K.E. and Basterndorf K.M.(1980). Damping and slip of nailed joints. *Wood Science* **12** (4).

Brock, G.R.(1968). The behavior of nailed joints under long-term and short duration loading. Int. Symp. on Joints in Timber Structures, London.

Feldborg, T. (1989) Timber joints in tension and nails in withdrawal under long-term loading and alternative humidity. *Forest Products Journal,* **39** (11/12), 8–12.

Feldborg, T. and Johansen, M.(1986). Slip in joints under long-term loading. CIB-IUFRO Meeting, Florence.

Foschi, R.O. (1974) Load slip characteristics of nails. *Wood Science* **7** (1).

Jang, S., and Polensek, A. (1989) *Theoretical models for creep slip of nailed joints between wood and wood-based materials.* Paper 2288. Forest Research Lab. Corvallis, Or. USA.

Jenkins, J.L., Polensek A. and Basterndorf K.M.(1979) Stiffness of nailed wall joints under short- and long-term lateral load. *Wood Science* **11** (3).

Kuipers, J. (1977). Long duration tests on timber joints. CIB-W18 Meeting, Stockholm.

Leicester, R.H., Reardon, G.F.,Schuster K.B.(1979). Toothed plate connector joints subjected to long duration loads. 19th Forest Products Research Conference, Melbourne.

Leitjen, A.J.M. (1988). Long duration strength of joints with high working load levels. Int. Conference of Timber Engineering, Seattle.

Leivo, M. (1989). Creep in wooden structures. RILEM Meeting, Helsinki.

Mack, J.J. (1963). A study of creep in nailed joints. CSIRO, Australia, Div. Forest Prod. Tech. Paper 27.

Noren, B. (1968). *Nailed joints – Their strength and rigidity under short-term and long-term loading.* Report 158B, Nat. Swedish Institute For Building Research, Stockholm.

Palka, L.C. (1981). *Effect of load duration upon timber fasteners (a selective literature review).* Report, Forintek Canada Corp., Vancouver.

Palka, L.C. (1988a). Effect of long-term constant loads on truss-plate joints in tension under ambient laboratory conditions. Int. Conference of Timber Engineering, Seattle.

Palka, L.C. (1988b). *Exploratory study of the short-term and long-term behavior of truss-plate joints in tension under ambient laboratory conditions.* Report, Forintek Canada Corp., Vancouver.

Palka, L.C. (1989a.) A review of creep in timber structures: models and data. RILEM TC 112 Meeting, London.

Palka, L.C. (1989b) *Review of truss-plate joints in tension under ambient laboratory conditions.* Report, Forintek Canada Corp, Vancouver.

Whale, L.R.J. (1988). *Deformation characteristics of nailed or bolted timber joints subjected to irregular short or medium term lateral loading.* Thesis, South Bank Polytechnic, London.

Whale, L.R.J. and Smith, I. (1989), A method for measuring the embedding characteristics of wood and wood based materials. *Materiaux et constructions*, **22**.

Wilkinson, T.L. (1988). *Duration of load on bolted joints: a pilot study.* Res. Pap. FPL RP 488, US Dept. of Agriculture, Madison, Wisconsin.

7

Creep of wooden structural components: testing methods

C. Le Govic

7.1 Introduction

The review of creep data in timber structures is based on a limited number of experiments using different test methods. Table 2.1 p. 21 gives a recent collection of creep data from structural wooden components.

From a modelling point of view, and for final engineering use, a standardization of creep testing is recommended. With the support of a few detailed tests some guidelines are presented, concerning:

- Specimens and materials
- Load history (level, duration,)
- Climatic conditions and internal variables
- Deflection measurements.

Here, answers are sought to the question: 'What do we want to measure and how do we do it?' within the framework of the 'fundamental aspects on creep in wood'. Creep tests are performed for two different purposes:

- to provide experimental data for a specified material (as Eurocode 5 needs); or
- to perform experiments for research on modelling the viscoelastic behaviour of wood.

7.2 Specimens and materials

For an industrial application, creep studies of lumber and glulam beams and other wood-based materials, are performed on random samples from 'normal production'. For European countries it is possible to use the EN grade classes (from EN 338 or Pr EN 518 for lumber, and EN TC 124.207 for glued laminated timber).

From a research point of view, stratified sampling may be preferable or necessary based on non-destructive tests (ultasonic velocity, etc.), such as density, elastic measurements, etc.

For all wood-based materials, information on density, species or wood species mix, grade, and (allowable) defects are needed.

For glued laminated timber the following informations may help the analysis: the kind of glue, the size and number of laminae.

For lumber beams, the angle between load and the grain plane is a relevant factor.

The number of test specimens or replicates evaluated within a specified set of experimental conditions is variable in the literature. For example, in Rouger *et al.* (1990) the number of replications is three, it is five in Srpcic and Moody (1988) and ten in Hoyle *et al.* (1985). In North America, the number of replicates generally range from nine to seventy-five (Palka 1989,1992).

7.3 Mechanical test

Four-point (or three-point) bending tests are most frequently performed for beams, as well as panels.

From a general point of view, the creep tests should be derived from static standard tests. The span/depth ratio may range from 12:1 (lumber) to 60:1 (panels). For lumber and glued laminated beams the European static standard ratio is 18:1.

For mechanical instability problems (lateral buckling) the ratio (width/length) should be at least equal to 1/8. For research purposes we can have a ratio less than 1/8; in this case it is necessary to prevent delayed lateral instability.

Fig. 7.1(a) gives an illustration of an experimental set-up for large specimens. Figure 7.1(b) from Madsen and Barrett (1976) and Fig. 7.1(c) from Le Govic (1991) may be used for small and medium specimens. These authors used an arm moment amplification.

7.4 Load history

Load levels during creep tests for structural design purposes should be as similar as possible to those specified by building codes. Two load levels are proposed for experimental purpose:

- a level of 15 MPa which corresponds to the lumber and glulam limit state stress design

$$\sigma_d = \frac{\sigma_k \, K_{mod}}{\gamma}$$

- a level of 5 MPa which corresponds to a normal constant stress for a wooden structure.

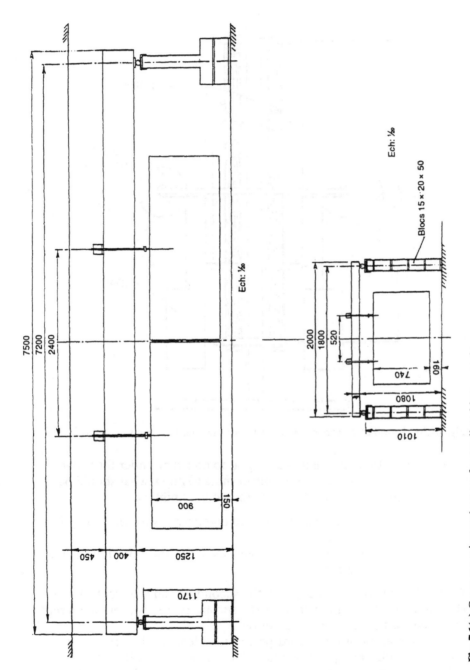

Fig. 7.1(a) Creep experimental set-up for small and large specimens.

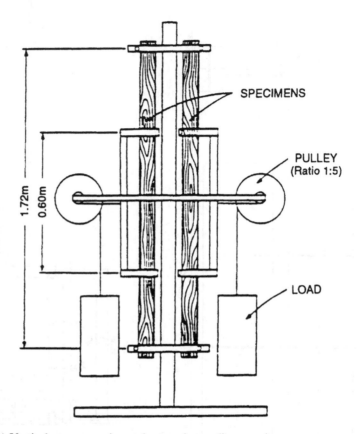

Fig. 7.1(b) Vertical creep experimental set-up for medium specimens.

The load duration may be at least equal to one year under these two constant load levels. The interesting point is to know what should be done after the creep period. There are two possibilities:

- determine the residual strength, and compare it with the strength of reference material, or
- induce rupture by step loading and evaluate the duration of load, with an adequate model.

After a loading period of 1 month the second period is equal to 1 week. Between 6 months and 18 months the second period is around 5 weeks increasing with the total creep period.

This increasing of the record period may lead to the use of manual recording tools like optical sights or hand-held mechanical sensors.

Note, however, that creep recordings at specified time intervals would result in storing identical deflection values for adjacent times. To minimize the computer storage needed for creep tests of long duration

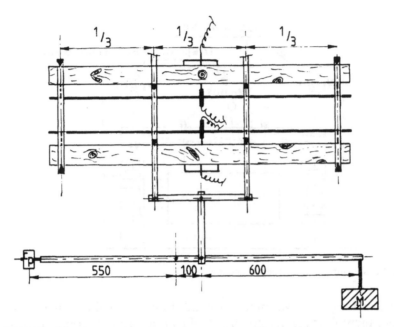

Fig. 7.1(c) Horizontal creep experimental set-up for medium specimens.

and as a high number of replicates, the data acquisition system was programmed to record creep only be specified increments, such as 0.025 mm in the first four months and 0.05 mm thereafter, in a truss-plate joint and waferboard study, see Palka (1989, 1992).

The measurement of relative deformation is an imperative requirement. The total absolute deflection W_T includes the creep transverse compression behaviour of wooden specimens (with local effects such as punching) W_1, the delayed behaviour of load frame supports (packing of concrete supports, for example) W_2 and the bending behaviour W. Figs. 7.2 and 7.3 illustrate these considerations with $W_T = W_1 + W_2 + W$.

The measurement apparatus used in Rouger *et al.* (1990) is presented in Fig. 7.2 with the corresponding creep curves presented in Fig. 7.3. In this case the model was fitted on a mix between.

1) the deflection due to bending behaviour with the punching effect for the first week measured with displacement transducers and,
2) the delayed bending behaviour measured with optical sights.

7.5 Climatic conditions and internal variables

It is generally accepted that the constant standard humidity environmental conditions are the 'air dry' (65% RH, 20°C) and the 'humid' (85% RH, 20°C).

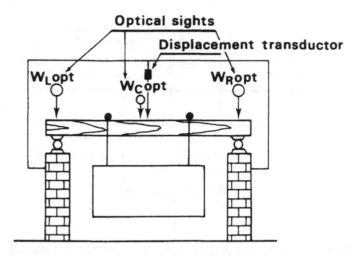

Fig. 7.2 Measurement of creep deflection.

The changing environmental conditions are defined in relation to experimental or service conditions. From a research point of view, the temperature and relative humidity are external parameters varying as mathematical functions of time (step or sinus with different frequencies or amplitudes).

From a code point of view, these conditions should correspond to protected 'internal' and 'external' conditions where temperature and relative humidity are variable (non-controlled) climatic inputs.

From a modelling point of view, it is necessary to know the spatial and time gradients of moisture content and temperature. A very interesting experimental approach was given in Srpcic and Moody (1988). They determined the time evolution of MC, from measurements made with a resistance type moisture and temperature meter at different points (five, for example) of the cross-section of slices cut from extra beams. Moisture content calibration was done by weight measurements. Temperature should always be measured because of the influence of non-controlled parameters: air movement, sun orientation in relation to protected material, etc.

7.6 Deflection measurements

The equipment used for the tests should present the following capabilities:

- record the early part of the creep with a good time resolution (1 second as a minimum time scan),

- ensure the stability of the measurement for a long period of time (several years), despite the possibility of geometric changes of the specimen,
- measure a relative deformation.

To achieve these requirements, different instrumentation may be used with overlapping periods.

In general the deflection is only measured either on the tensile or compressive edge of the beam. Hoffmeyer (1991) has recorded the creep behaviour from tension and compression edges.

Recording the beginning of wood creep phenomena needs an automatic measurement of the deflection. The following time recording sequence was used in Rouger *et al.* (1990):

- each ten seconds between 0 and 2 hours
- each thirty seconds between 2 and 16 hours
- each minute between 16 hours and 2 days
- each thirty minutes between 2 and 7 days

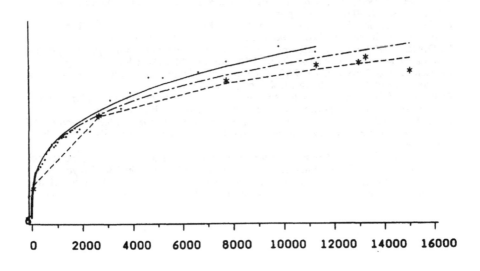

MEASUREMENTS
- .... W^{trans} + W1 (measured on the short period with displacement transductor).
- * * * * W^{opt} (measured on all the creep period with optical sights).

CREEP FUNCTION
- _____ f^{trans}: based on (W^{trans} + W1) measurements.
- - - - - f^{opt}: based on (W^{opt}) measurements.
- - . - . - f^{mod}: mix between f^{trans} and f^{opt}.

Fig. 7.3 Superposition of creep measurement and modelling

Fig. 7.1(c) gives another possibility of measuring the relative deflection during wood bending creep either including or ignoring shear behaviour.

The sensor (optical or electronic) shall be mounted on a rigid support, which is not easy to make for very long beams. Alternatively, the sensors may be mounted on the shorter 'shear free' segment of three-point loaded specimens (Palka 1992).

7.7 Conclusion

As a conclusion we would recommend the reading of the first European standard dealing with creep of wood-based panels which is written under the authority of J. Dinwoodie: *'Determination of Duration of Load and Creep Factors'*, EN TC112 (DOC N129, 1992).

7.8 References

Hoffmeyer, P. (1991). *Failure of wood as influenced by moisture and load duration.* EEC contract MA 1B-0042; Final report.

Hoyle, R.J., Griffith, M.C. and Itani R.Y. (1985). Primary creep in Douglas-fir beams of commercial size and quality. *Wood and Fiber Science,* **17** (3), 300–314.

Le Govic, C. (1991). *Creep and time to failure in structural timber with special reference to the influence of climatic conditions.* EEC contract MA 1B-0040F; Final report.

Madsen, B. and Barrett, J.D. (1976). *Time-strength relationship for lumber.* Struct. Res. Series, Report 13, Univ. British Columbia, Vancouver.

Palka, L.C. (1989). Review of truss-plate joints in long term tension under ambient laboratory conditions. Forintek Canada Corp. Vancouver, B.C., 84 pages.

Palka, L.C. (1992). *Long term behaviour of waferboard panels in bending: experimental procedures and residual strength properties.* Forintek Canada Corp. Vancouver, B.C., 95 pages.

Rouger, F., Le Govic, C., Crubilé P., Soubret, R. and Paquet, J. (1990). Creep behaviour of French woods. *Timber Engineering Conference,* Vol. 2, pp. 330–336, Tokyo.

Srpcic, J. and Moody, R.C. (1988). Creep of small glulam beams under changing relative humidity conditions. *International Conference on Timber Engineering,* pp. 523–530.

8
Variation of moisture content in timber structures

S. Thelandersson

8.1 Introduction

It is well known that creep in timber structures depends on the moisture content *level*. It is also well known from laboratory tests as well as field tests that creep in wood is strongly affected by *variation* in moisture content. Therefore, predictions of creep in timber structures in practice must be based on knowledge of the actual moisture content and its variation in the material.

Building components in practice are almost always exposed to changing climatic conditions of a more or less random nature. The moisture content u depends generally on both position r within the timber element and time t, i.e., u = u(r,t). A detailed description of u(r,t) is not feasible in most practical situations. Instead it is desirable to identify general trends of the behaviour related to specific types of climate exposure.

A common way to simplify the problem is to study the average moisture content $u_m(t)$ within a timber section, i.e.

$$u_m(t) = \frac{1}{A} \int_A u(r,t)\, dA$$

Most data found in the literature of moisture conditions in timber elements refer to such average values. But for the present purpose it is also important to have information about the non-uniform moisture distribution within the section, as well as the nature of time variations of moisture content at different depths.

8.2 Climatic exposure

Climatic exposure of structural timber elements may be classified as follows:

1) Direct weathering exposure
2) Weathering exposure under shelter
3) Structural element in climate separating components

4) Structural element in non-heated spaces
5) Indoors climate.

Cases 2 and 5 are suitable as reference cases for an estimation of moisture content variation in building timber (Hanson 1987, 1988). They can be seen as limit cases for the moisture exposure of most building components including those of cases 3 and 4 above. Case 1, with the timber exposed to rain and snow is the more extreme and infrequent situation.

Tsoumis (1964) has estimated the moisture content (MC) in timber based on meteorological data for different locations in Europe. Some results are shown in Table 8.1. His MC values were based on the assumption that the timber is in moisture equilibrium with the monthly average of relative humidity. For timber of relatively small dimensions this may be a reasonable estimate of the average moisture content corresponding to case 2 above.

Table 8.1 Yearly maximum and minimum moisture content in timber (outdoors under shelter) (Tsoumis 1964; Hanson 1988)

Location	Maximum %	Minimum %
Stockholm	20	14
Helsinki	20	14
Oslo	18	14
London	20	14
Paris	20	14
Berlin	20	14
Madrid	16	8
Rome	14	12
Cairo	14	12

The moisture conditions inside inhabited buildings (case 5) depends on the climate conditions outside, the inside temperature, the production of moisture within the building and the rate of ventilation. Fig. 8.1 shows the variation over one year of relative humidity indoors for three different places in Sweden (Nevander and Elmarsson 1961).

Both the external and internal climate has a general periodic variation over the year such as those described above. Superimposed on these annual cycles there are short-term fluctuations over periods lasting one or two weeks or a few days as well as daily cycles. This is especially true for the external climate. Significant short-term variations in internal moisture conditions may occur due to, for example, a temporary increase in moisture production from cooking, washing and bathing.

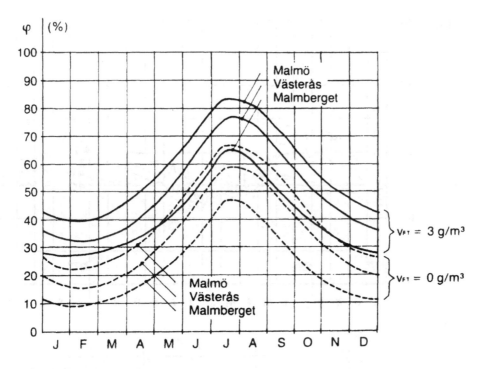

Fig. 8.1 Relative humidity indoors in southern Sweden (Malmö), central Sweden (Västeras) and northern Sweden (Malmberget) for two different levels of internal moisture production V_{FT} (Nevander and Elmarsson 1961).

8.3 Long-term variation of MC

The response of timber sections to variation in the surrounding climate depends on:

- size and shape of the member
- the tightness of surface coating, if any
- materials properties related to moisture transport and moisture fixation
- moisture transfer conditions at the boundaries.

Meierhofer and Sell (1978–9) performed comprehensive experimental investigations of MC variations in timber beams exposed to natural 'weathering under shelter'. Fig. 8.2(a) shows the time variation over 21 months of u_m for beams with characteristic dimensions ranging from 10 mm to 170 mm. Results from the same investigation are also presented in Fig 8.2(b) in the form of cumulative distribution of MC over a period of one year.

As expected, there is a greater variation for smaller dimensions: in

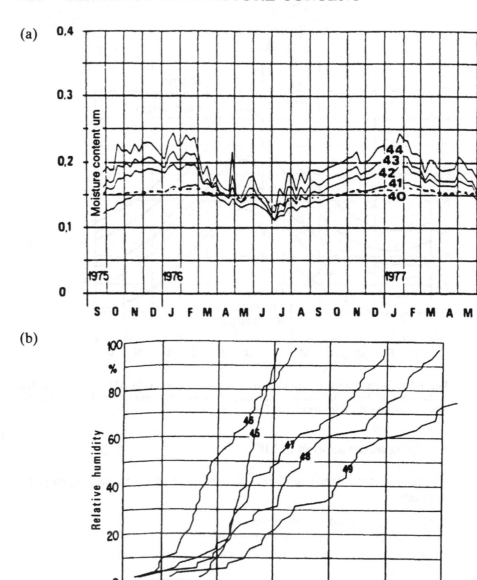

Fig. 8.2 (a) Time variation of u_m (average MC in the timber section) from Sept. 1975 to May 1977 (Meierhofer and Sell 1978–9).
(b) Cumulative distribution of MC during one year (Meierhofer and Sell 1978–9).
Softwood beams of different sizes (B × H mm).
40, 45 : 170 × 340
41, 46 : 80 × 160
42, 47 : 40 × 80
43, 48 : 20 × 40
44, 49 : 10 × 20

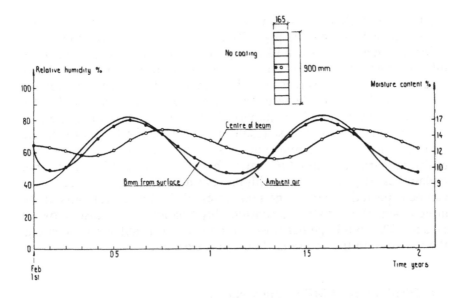

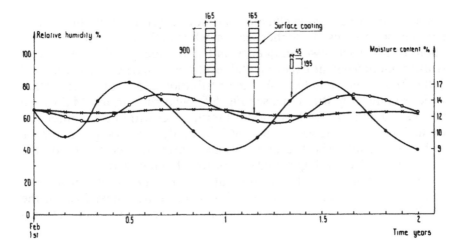

Fig. 8.3 (a) Variation of MC in different points of a large glulam beam section exposed to indoor climate typical for southern Scandinavia.
(b) MC in the centre of beams of different sizes. The moisture resistance for the surface coating was taken to 5×10^6 s/m.

the 10 mm beam MC values ranging from 12% to 24% were observed, while in the largest 170 mm beams the range is only between 13% and 16%. The frequency of occurrence over the year is nearly uniform (the cumulative distribution curve is approximately linear). It is more surprising, however, that the mean MC over the whole time period is significantly higher for beams of smaller dimensions. Beams with smaller dimensions also respond faster to short time fluctuations.

Measurements of MC timber exposed to indoor climate are more difficult to find in literature. But since the temperature in this case is fairly constant and the exposure is much less random, theoretical analyses can be made with reasonable accuracy, see, for example Arvidsson (1989). As an example, results from numerical analyses are shown in Fig. 8.3.

The conclusions from these results are that structural timber with the typical dimension of 45 mm is in quasi-equilibrium with the long term moisture variation, and MC varies between 9 and 17%. For large glulam beams (b = 165 mm) without coating, MC in the surface layers can be expected to vary in the small interval. In the central parts of the large beams the variation is considerably smaller, in this case between 11 and 15%. In a large beam with a moisture-resistant surface coating, the variation in the centre is almost non-existent.

8.4   Distribution of MC within sections

The response to changing climate conditions is always more immediate near the boundaries of a timber section, while in the inner parts the variations of MC have lower frequency and amplitude, see Fig. 8.3(a). This is also evident from the experimental results obtained by Meierhofer and Sell (1978–9) shown in Fig. 8.4. The distribution of MC over the section varies significantly with time, see Fig. 8.5. Near the surface, MC varied between 11 and 19% with many short term cycles within 2–3% intervals. In the inner parts the variation of MC was only in the range 12 to 16% with much smaller short-term variations. During the winter season the distribution of MC is highly non-uniform within the section.

Structural timber in climate separating components also have non-uniform MC especially in winter, with significantly higher MC on the 'cold' side. This can lead to large moisture-induced deflections of such components.

8.5   Short-term variations

It is reasonable to assume that short-term local variations of MC has significant influence on the structural behaviour only if they are active in a zone of say 10 mm or more near the surface. It can be concluded from Figs. 8.2(a) and 8.4(a) that the response at a depth of the order of 10 mm corresponds to humidity variations in the surroundings with a period of one week or more. Daily variations, which can be of large magnitude during some periods of the year, are too short to give a

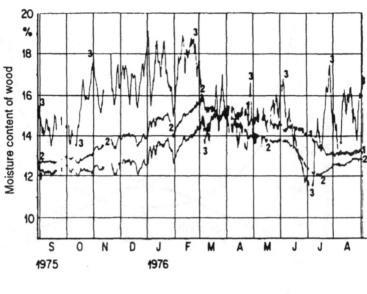

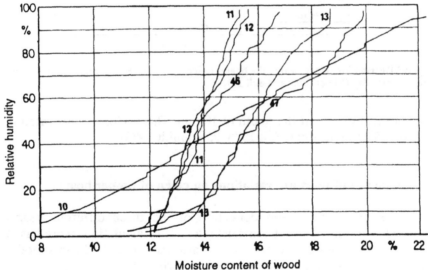

Fig. 8.4 Moisture conditions in a softwood timber section (170 × 340 mm) outdoors under shelter (Switzerland) Meierhofer and Sell 1978–9).
a) Time variation of MC at different points of the section.
b) Cumulative distribution of MC during one year.
    1, 11: Centre of section.
    2, 12: 40 mm from surface.
    3, 13: 10 mm from surface.
    10: MC corresponding to equilibrium with the air.
    46, 47: The same as in Fig. 8.2(b).

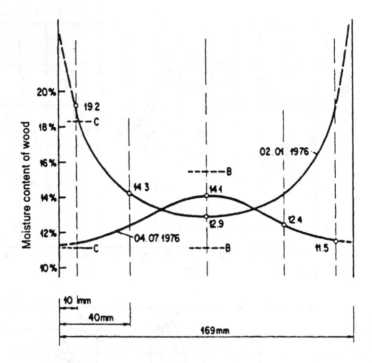

Fig. 8.5 Distribution of MC within the beams described in Fig. 8.4 for one winter day and one summer day (Meierhofer and Sell 1978–9).

significant response at the depth of 10 mm. This is clearly illustrated in Fig. 8.6 (Meierhofer and Sell 1978–9; Schmidt 1986).

8.6 Influence of moisture variations on creep of timber beams

As an example, the creep deflection of the three different beams referred to in Fig. 8.3 has been calculated. The calculations were based on the material model described in Märtensson (1992). In this model the mechano-sorptive strain rate ε_{ms} is described by:

$$\dot{\varepsilon}_{ms} = \kappa \frac{\sigma}{E_{ref}} \mid \alpha \dot{\mu} \mid$$

where:

σ = stress
$\dot{\mu}$ = moisture content
α = shrinkage / swelling coefficient
E_{ref} = elastic modulus at w = 20%
κ = $\kappa(\sigma)$ = material parameter.

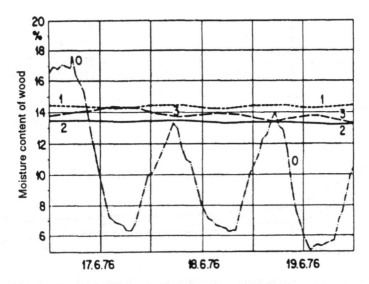

Fig. 8.6 Time variation of MC at points 1–3 (see Fig. 8.4(a) during three consecutive summer days. Curve 0 shows the MC corresponding to instant equilibrium with the relative humidity of the surrounding air (Meierhofer and Sell 1978–9).

The value of κ depends on moisture history and in the calculations presented here, κ is taken to:

$$\kappa = \begin{cases} 700 \text{ if } \sigma > 0 \\ 800 \text{ if } \sigma < 0 \end{cases}$$

for the first moisture change. For subsequent moisture changes κ is assumed to be zero. Also $E_{ref} = 10\,400$ MPa , $\alpha = 0.016$ % / %MC.

The results of the calculations are shown in Fig. 8.7. As expected, the creep deflection is largest for the small timber beam (45×195), but the creep of the large glulam beam (215×1260) is only slightly smaller. In the large beam protected by paint, the creep deflection is significantly smaller than in the unprotected one. The results show clearly the practical significance of moisture variations on creep.

8.7 Summary and conclusions

The variations of moisture content (MC) in structural timber to be expected in practice is of a very complex nature. The following conclusions may be appropriate

1) The variation over the year of average MC for timber of moderate sizes roughly corresponds to quasi-equilibrium with monthly mean values in relative humidity.

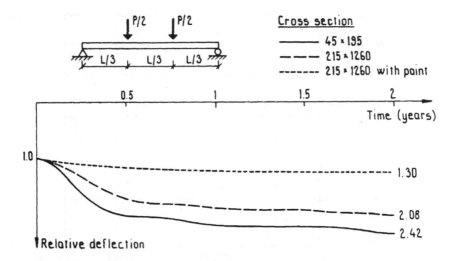

Fig. 8.7 Calculated creep deflection for beams with different sizes. For moisture variation, see Fig 3.

2) Timber with large sizes does not respond fully to climate cycles over the year.
3) The variation of MC in zones near the surfaces is much greater and more frequent than in the inner parts of a timber section.
4) Temporary climate changes with a duration of only one or two days has little influence on the structural behaviour of timber structures.
5) In most cases structural timber in building is exposed to humidity conditions in between the two extremes of outdoor climate and indoor climate.
6) If moisture is considered as important in connection with structural safety and serviceability, a probabilistic, reliability-based approach should be employed similar to that used for others loads.

8.8 References

Arvidsson, J. (1989). *Computer model for two-dimensional moisture transport*. Manual for JAM-2. Dep. of Building Physics, Lund University, Lund, Sweden.

Hanson, T. (1987). *Fuktvot i inbyggt virke* (Moisture Content in Built-in Timber). Träteknikcentrum, Report I 8706038, Stockholm (in Swedish).

Hanson, T. (1988). *Fukt for exporttä - Resultat av förstudie*. Träteknikcentrum, Report I 8802006, Stockholm (in Swedish).

Märtensson, A. (1992). *Mechanical behaviour of wood exposed to humidity variations*. Report TVBK – 1006, Lund Institute of Technology, Sweden.

Meierhofer, U., and Sell, J. (1978-79). Physikalische Vorgänge in Wetterbeanspruchten Holzbauteilen. *Holz als Roh- und Werkstoff*, 1. Mitt., **36**, 461–466, 2 Mitt., **37**, 227–234, 3. Mitt., **37**, 447–454.

Nevander, L E. and Elmarsson, B. (1961). *Fukthandboken*, Svensk Byggtjänst, Stockholm (in Swedish).

Schmidt, K. (1986). Untersuchungen über die Raumklimapuffering durch Holzoberflächen. *Holzforschung und Holzverwertung* **38**, 1.

Tsoumis, G. (1964). Estimated moisture content of air dry wood exposed to the atmosphere under shelter, especially in Europe. *Holzforschung* **18**.

9
Creep analysis of timber structures

A. Ranta-Maunus

9.1 Code method

In timber structures creep is most prominent in the deflection of beams. Simplified methods are used to calculate the long-term deflection based on the modulus of elasticity E and modified for the type of loading. Numerical values of the correction factors in different countries are given in the appendix. Simplified 'pseudoelastic' design methods are under development for the design of wood and panel structures (Palka 1991).

In Eurocode 5, the following method is adopted. Values of E are given in material standards at a reference moisture content. Elastic deflection u_{inst} is calculated by using those values.

Then, the long-term deflection is calculated by the expression:

$$u = u_{inst}(1 + k_{def})$$

(9.1)

where k_{def} takes into account both the creep, depending on the load duration, and the change of E at higher moisture contents. Values of k_{def} are given in tables for different wooden materials. When the deflection consists of the contribution of bending stresses, shear stresses and joint slip, all these components have to be taken into account:

$$u = \sum_i u_{inst,i}(1 + k_{def,i})$$

(9.2)

where i = 1 refers to bending, i = 2 to shear, i = 3 to slip.

Equation (9.2) is for one type of loading only. When calculating the maximum total deflection, different loads and precamber of the beam are simultaneously considered as follows:

$$u = u_0 + \sum_j \sum_i u_{inst,i,j}(1 + k_{def,i,j})$$

(9.3)

where u_o is the precamber of the beam (negative) and subscript j refers to different load duration classes (permanent, medium, etc).

Note that the k_{def} values in Eurocode 5 are only guidelines. Other values can be used if more accurate information is available for specific cases.

Accurate determination of k_{def} values for variable loads is difficult. This is because the design values of loads are the maximum which the structure has to resist. For instance, the design value of snow load is expected to be present only once in 50 years; and the peak value may last a very short time. u_{inst} is calculated based on such an unusual value. The design of columns in Eurocode 5 is not related to creep properties.

9.2 Linear viscoelastic analysis

The creep under changing loads has been traditionally analysed by the use of the linear theory of viscoelasticity. The practical limitation is that wood only follows the theory relatively well under constant climatic conditions.The constitutive equation

$$\varepsilon = \int_0^t J(t - \tau)d\sigma(\tau)$$

$$(9.4)$$

is based on the Boltzmann principle of superposition. It is a convenient mathematical formulation, for which simple analytical solutions can also be found.

Equivalent formulations could be made by the use of differential equations, which are often more suitable for numerical solutions.

It has been shown that the stress distribution under a sustained load does not change in a structure composed of materials with the same linear creep function. The main application of linear viscoelastic analysis are:

• analysis of displacements under varying loads under constant environmental conditions
• analysis of stresses and displacements in structures composed of different materials.

A step-by-step matrix method for the frame analysis has been presented by Tardos *et al.* (1977). It has been applied to continuous wooden beams with mechanical joints at supports (Capretti *et al.* 1988). When a joint has a higher creep rate than timber, the bending moment at the support will decrease, and the moment at mid-span will increase, with time. Linear viscoelastic methods have also been applied to changing climatic conditions, simply by taking the creep functions based on experiments under similar conditions.

Creep buckling of columns is the most obvious application of creep analysis to design for the ultimate limit state. The analysis is based on the calculation of the increase of induced deflection towards the final critical value. The effect of load is non-linear, because the bending moment is increasing with increasing lateral deflection. Numerical procedures have been applied to solve problems of this type (Itani *et al.* 1986; Blass 1988).

9.3 Mechano-sorptive creep analysis

The effect of varying moisture content on beams under constant load can be easily analysed, when the cyclic variation of the average moisture content is taken into account, and the moisture gradients are neglected.

Creep deformation induced by varying load and moisture content can be analysed when a mechano-sorptive constitutive equation is available. Based on the equation:

$$\left\{ \dot{\varepsilon}^{ms} \right\} = \left[S^{ms} \right] \left\{ \dot{\sigma} \right\} u$$

$$(9.5)$$

for mechano-sorptive strain rate, and omitting viscoelastic strain for simplicity, a 3-D method for stress analysis was developed in the ABAQUS FEM-program (Santaoja *et al.* 1991). The method was applied to the analysis of the effect of a sinusoidal varying relative humidity (Hanhijärvi and Ranta-Maunas 1990). The resulting deflection of a beam is illustrated in Fig. 9.1.

A combined diffusion and creep analysis program has also been used for the flexural analysis of beams (Toratti 1991). A two-dimensional diffusion model calculates the values of moisture content at different locations of the cross-section at various times. In turn, these are used as initial values in structural analysis, which is based on the beam theory with Bernoulli assumption. Different constitutive equations are used involving elastic, viscoelastic, mechano-sorptive and shrinkage elements. Mainly various test situations are analysed. The method has been applied also to demonstrate the influence of cross-section size and moisture cycle length to the mechano-sorptive creep deflection.

A simulation of the deflection of roof structures loaded partly by random snow load and partly by dead load has been made (Toratti 1992). The simulation is based on information of the outdoor relative humidity in Helsinki, and on the statistics of snow depth in Scandinavia. The result is illustrated in Fig. 9.2 where the time-dependent deflection is shown relative to the elastic deflection caused by the design load. The design value of snow load corresponds to a 50-year return period. The annual values of simulated snow load vary between 25% and 96% of the design value.

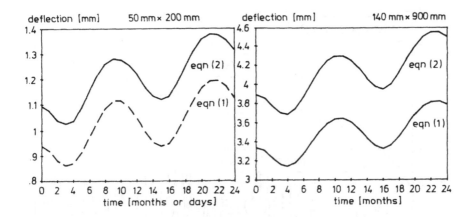

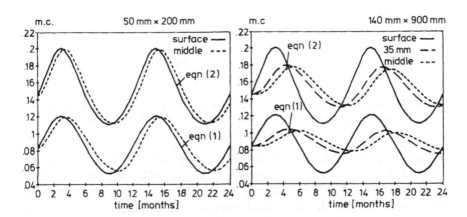

Fig. 9.1 Computed deflections and local moisture contents in timber based on an annual sinusoidal humidity cycle (Ranta-Maunas 1991).

9.4 Residual stresses and long-term strength of wood

Creep affects the stress distribution in timber cross-sections in two ways:

- drying of wood and moisture changes cause stresses perpendicular to the grain, which may cause checking of the initially stress-free wood, and
- cyclically varying moisture content in a loaded wooden member causes a redistribution of stresses, which may be related to the duration of load phenomenon.

During drying of timber, stresses perpendicular to the grain appear due

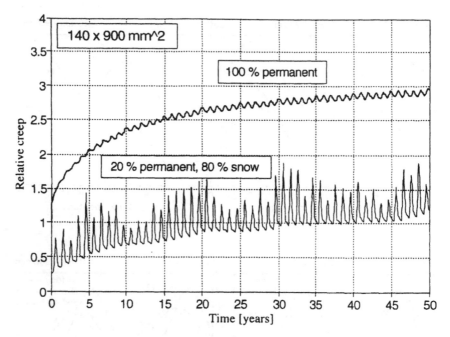

Fig. 9.2 Computed relative deflection of 140 × 900 mm² glulam beam for permanent load and for a combination with a random snow load (Toratti 1992).

to moisture gradient and shrinkage properties of the material. In the past, the drying stresses were first calculated by the use of purely elastic methods. This predicted stresses several times higher than the strength of wood in the tangential direction. More realistic values were obtained by plastic analysis. This is, however, the wrong method for wood under tension, when the behaviour is far from being plastic. Viscoelastic analysis was the next step in developing methods for the estimation of drying stresses. This method also gives too high values for stresses. The first mechano-sorptive analysis of drying stresses was made by Takemura (1973). Recently, several researchers have been active in this field (Ranta-Maunus 1989; Felix 1990; Thelandersson and Moren 1990, Salin 1990). Fig. 9.3 shows an example of the development of stresses at the surface and in the middle of sawn timber during drying.

Nielsen (1982) has presented a theory for the long-term strength of wood based on a combination of fracture mechanics and viscoelasticity. He considered wood as a linear viscoelastic material containing initial cracks. Then, lifetime was determined by the critical size of gradually growing cracks. A review of his work and those of others is given by Valentin (1991) in a RILEM state-of-the-art report: *Application of Fracture Mechanics to Timber Structures*.

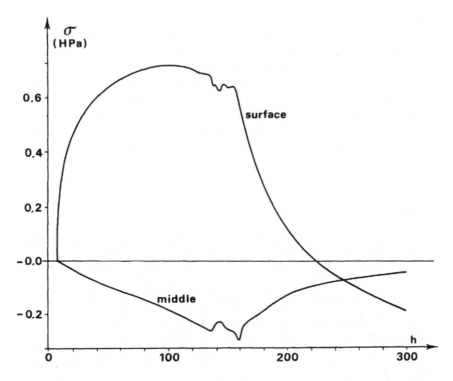

Fig. 9.3 Illustration of drying stresses at the surface and in the middle of sawn timber.

A simple one-dimensional mechano-sorptive analysis shows that the stress distribution of wooden members is changing with time due to different moisture levels and variation at the surface and in the middle (Ranta-Maunus 1990). This leads to the accumulation of stresses which may be related to the duration of load phenomenon.

9.5 Research needs

A structural analysis requires :

- information about load sequences as well as environmental (moisture and temperature) history,
- a constitutive equation of the material(s), and
- computational methods for solving the problem.

All these three areas have to be covered before a practical problem can be solved. Computational structural analysis methods are, in principle, the same for different applications. Furthermore, constitutive equations are basically independent of the problem, except that these have to be known at different orientations and in various temperature and moisture regimes. Load and service conditions vary depending on the case: snow on the roof, occupancy on the floor, pre-tension in a bridge, or high-temperature drying of sawn timber.

There is clearly a lack of resources for examining all areas, from loads to computations, in great detail. Because constitutive equations are discussed elsewhere, the importance of knowing and modelling the loads and environmental conditions is emphasized here. This includes the diffusion analyses both for timber drying and for structures in service.

9.6 References

Blass, H. J. (1988). The influence of creep and duration of load on the design of timber columns. *Proceedings of the 1988 ICOTE*, Vol. 1, pp. 365–373, USA.

Capretti, S., Ceccotti, A., and Spinelli P. (1988). Glued laminated timber hyperstatic structures: influence of creep under variable thermo hygrometrical conditions. *Conference on 'Comportement Mechanique du Bois'* Bordeaux, France pp. 423–432.

Felix, S. (1990). Modelisation des deformations et contraintes dans une piece de bois en cours de séchage. Thesis no. 414, University of Bordeaux, France.

Hanhijärvi, A. and Ranta-Maunus, A. (1990). A three-dimensional analysis of wooden beams under changing humidity conditions. IUFRO S5.02 meeting, Saint John, Canada. 10 p.

Itani, R.Y., Griffith, M.C., and Hoyle, R.J.Jr. (1986). The effect of creep on long wood column design and performance. *ASCE Journal of Structural Engineering* **112**(5), 1097–1114.

Kawai, S., Nakato, K. and Sadoh, T. (1979). Computation of drying stresses resulting from moisture gradients in wood during drying I. Computative method. *Mokuzai Gakkaishi* **25**,(2), 103–110.

Morlier, P. (1990). Time dependent analysis of timber structures. IUFRO 5.02 meeting, Canada.

Nielsen, L. F. (1982). A life time analysis of cracked linear viscoelastic materials with special reference to wood. IUFRO Timber Engineering Meeting, Sweden. pp. 153–178.

Nielsen, L. F. (1991). Lifetime, residual strength, and quality of wood and other visco-elastic building materials. *Holz als Roh- und Werkstoff* **49**, 451–455.

Palka, L. C. (1991). *Simplified engineering approach to creep and creep rupture for wood and wood based materials.* Forintek Report. Forintek Canada Corp. Vancouver. 31 p.

Ranta-Maunus, A. (1986). Drying stresses in round timber. CIB-W18/IUFRO S 5.02-meeting, Italy. 11p.

Ranta-Maunus, A. (1989). Analysis of drying stresses in timber. *Paper and Timber* **10**.

Ranta-Maunus, A. (1990). Impact of mechano-sorptive creep on the long-term strength of timber. *Holz als Roh- und Werkstoff* **48**, 67–71.

Razafindrakoto, J.-C., and Valentin, G. (1984). Rheological problems in models of soft-wood drying (in French). Rheology of anisotropic materials CR 19 Coll. Paris, pp. 543–560.

Salin, J.-G. (1990). Simulation of the timber drying process. Prediction of moisture and quality changes. Dissertation in Åbo Akademi. EKONO Oy, Helsinki.

Santaoja, T., Leino, T., Ranta-Maunus, A., and Hanhijärvi, A. (1991). *Mechano-sorptive structural analysis of wood by ABAQUS finite element program.* Technical Research Centre of Finland, Research Notes 1276. 34 p. + app. 16 p.

Takemura, T. (1973). The memory effect of wood and the stress development during drying (in Japanese). *J. Soc. Mat. Science Japan* **22,** 236, 476–478.

Tardos, M.K., Ghali, A., and Dilger, W.H. (1977). Time dependent analysis of composite frames, *ASCE Journal of Structural Division*, **21**.

Toratti, T. (1991). Creep of different size wood members in varying environment humidity. *Proceedings of ICOTE*, London, pp.4.239–4.246.

Toratti, T. (1992). Modelling the creep of timber beams. *Journal of Structural Mechanics* (Rakenteiden Mekaniikka lehti). **25**(1), 12–35.

Thelandersson, S., and Moren, T. (1990). Tensile stresses and cracking in drying timber. IUFRO S 5.02 meeting, Canada. 29 p.

Valentin, G.H., Boström, L., Gustafsson, P.J., Ranta-Maunus, A. and Gowda, S. (1991). *Application of fracture mechanics to timber structures*, RILEM state-of-the-art report. VTT Research Notes 1262. Technical Research Centre of Finland, Espoo. 142 p. + app. 2 p.

Welling, J. (1988). Die modellmässige Erfassung von Trocknungsspannungen während der Kammertrocknung von Schnittholz. *Holz als Roh- und Werkstoff* **46**, 295–300.

10
Conclusion: priority research needs

P. Morlier

At the end of this survey, we can propose priority research in the following areas.

10.1 Fundamental

The basic mechanisms for creep in wooden materials need to be clarified:

1. explanation by chemical, molecular (ultra, micro) structural approach,
2. better identification of transitions ($1, 2, \varepsilon_1, \varepsilon_2, \alpha\beta$) and of the time–temperature domain where they occur,
3. better definition of local (ultra structural level) moisture content and temperature,
4. mechanisms of wood stabilization (acetylation, formaldehyde crosslinking).

For a better understanding, but also to provide data for Point 3, the experimental field has to extended as follows:

- tension, compression, shear,
- along the grain, across the grain,
- influence of ultra and micro structure on the properties.

We need for this consistent and greater cooperation between laboratories.

One of the most important thing is to confirm (or not) the existence of creep limit.

10.2 Characterization

From the references quoted in our report, we have obtained very scattered data without any logical classification. For this reason we need recommendations for creep measurement on wooden materials (clear wood, boards, lumber, joints, etc.) in order to improve the basic knowledge.

For example, for clear wood, those recommendations have to consider:

- size, shape of samples, type of loading (tension, compression, shear),
- structural properties of wood,
- selection, twinning of samples,
- conditioning (thermo-hygrorecovery may drastically reduce creep),
- environmental conditions (temperature, moisture content) and control of them,
- strain measurements,
- analysis of experimental data (how to define the instantaneous deformation),
- modelling.

A new Technical Committee of RILEM (TC 155) has just been created for this purpose.

10.3 Constitutive equations

These have to describe the relation between deformation, stress, temperature, moisture content and their first derivatives (taking into account the previous history of the material) at two levels:

1D (force-displacement), for a given experiment,
- tension of a sample,
- bending of a beam,
- tension of a joint, etc.
2D, 3D for the elementary volume and for future structural analysis.

They have to be tested on complex paths (loading-unloading; temperature, moisture content or relative humidity changes) provided we have a set of reliable experiments. The principal aims are:

- strain recovery,
- existence of creep limit,
- relaxation,

Their predictive character is in great demand.

10.4 Structural analysis

The constitutive equations should be applied in structural analysis together with heat and mass transfer models in order to simulate stresses and strains in structures.

The first difficulty is to simulate moisture content distribution in timber elements (beams, joints): do we need a 2D or a 3D model (role of the diffusivity anisotropy)?

The mechanical modelling may be simple (classical strength of materials) or more sophisticated (finite elements, etc.): as soon as we study curved or short beams we need at least a 2D model and the necessary 2D constitutive equations. The principal aims are:

- size effect,
- crack growth,
- stresses and strains due to mechano-sorptive creep in beams;
- the building up of consistent international cooperation.

10.5 Timber construction

10.5.1 From clear wood to timber

Structural timber involves wood with singularities (knots, sloping grains, juvenile wood); their influence on creep and on mechano-sorptive creep is not known. Questions concerning selection and grading of timber for design against mechano-sorptive creep, and reliability-based design are asked by D. Hunt; stabilization of timber by coatings or more sophisticated protections must be studied.

10.5.2 Mechanical joints

We know that very heavy experimental campaigns have been achieved in different countries and that Whale's thesis was an important step towards a good understanding of the mechanical behaviour of such joints. Nevertheless we think a greater effort towards the following is necessary:

- modelling of joints (in tension, in bending, simple embedment tests) by using the new rheological knowledge on wood (plastic behaviour in compression, crack propagation, etc.);
- experimental testing of joints with variations of relative humidity; temperature; fatigue testing; associated modelling;
- recommendations for creep measurement (see point 2).

10.5.3 Design rules for glulam and glued joints deserve special attention

After A. Ranta-Maunus, simple creep models for design rules should be developed; this would include knowledge about:

- variation of loads and moisture content in structures,

- creep at low stress levels,
- recovery during partly unloaded periods.

10.6 References

Finally, we strongly recommend the reading of the following recent literature:

Fundamental Aspects on Creep in Wood, A Workshop organised by COST 508-Wood Mechanics (Lund, March 1991). Proceedings edited by the Commission of the European Communities.

Mechanical behaviour of wood exposed to humidity variations, Annika Martensson's doctoral dissertation, Lund Institute of Technology, Report TVBK-1006, 1992.

Comportement différé du materiau bois dans le plan transverse sous des conditions hydriques évolutives, Patrick Joyet's doctoral dissertation, University of Bordeaux, No. 812, 1992

Creep of timber beams in a variable environment, Tomi Toratti's doctoral dissertation, Helsinki University of Technology, 1992.

Appendix: National building code creep factors

C. Le Govic

A.1 Introduction

To take account of the evolution of the viscoelastic behaviour of wood and wood-based materials in terms of deformation in service, some wood design codes have introduced creep modification factors defined by:

$$J(t) = J(t_{ref}) (1 + k_{creep}) = J(t_{ref}) k_{def} \qquad (A.1)$$

where k_{def} is a function of duration of load, stress level, temperature (T), and moisture content (MC) history.

Note that k_{def} also depends on the reference time (t_{ref}).

For these reasons, the proposed load duration and moisture classes are presented before the creep factor tables.

A.2 Comparison of national codes

Germany, (Structural Use of Timber – Design and construction – DIN 1052 – Part 1; April 1988), the United Kingdom (Structural Use of Timber – Code of practice for permissible stress design, materials and workmanship – BS 5268: Part 2: 1991), Canada (Engineering Design in Wood (Limit States Design), CAN/CSA-086.1-M89) and the United States of America (Specification for Engineered Wood Construction, Load and resistance factor design – July 1991 (not for publication)) have no creep factor tables in their wood design codes. In the German code, an analysis is required if the self-weight, g, exceeds 50% of the overall load q. For wood members or plywood, the creep factor, φ, is given in (A.2).

$$1 + \varphi = [1/ (3/2 - g/q)] \text{ for MC} < 18\% \qquad (A.\ 2a)$$

$$1 + \varphi = [1/ (5/3 - 4g/3q)] \text{ for MC} > 18\% \qquad (A.\ 2b)$$

In the Canadian code, in lieu of a more accurate estimate, an upper limit deflection of (1/360) of the span, is imposed on the elastic deflection of all structural components. In the United Kingdom, the ratio is equal to 0.003 of the span; for a longer span domestic floor, the deflection under design load should not exceed 14 mm.

In the new American design code (private communication, not for publication). Table 10.1.1 gives different deflection limit functions of the construction type and the nature of the load.

A.2.1 France – Regles C.B. 71, de calcul et de conception des charpentes en bois, juin 1984.

Creep is taken into account only in bending. Solid wood and glued laminated timber are treated identically.

The French approach has different assumptions:

- a non-linear evolution of creep factors with stress as presented in Fig. A. 1,
- a lack of load duration classes,
- a dependence of creep factors on H which is the difference between the initial moisture content of the timber during construction and after the apparent stabilization of deformation (drying under load),
- a reference to a standard of 15% MC with a complex formula (A. 3) providing creep estimates for other values of MC.

$$k_{creep}(H, \Delta H) = \left[\frac{\frac{H + \Delta H}{12}}{1 - \frac{\Delta H - 5}{15 + \Delta H}} \times \frac{\sigma_{r\infty} - 0,2\ \overline{\sigma}_f}{\overline{\sigma}_f} \right]$$

$$(A. 3)$$

Moreover there exists a limitation of deflection with an upper limit with different ratios (1/300, 1/400, 1/500) depending on the purpose of the building.

A.2.2 Finland – Finnish design code for wooden structures RIL 120–1986.

Three load duration classes and four moisture content classes are specified in table A. 1.

Table A.1

Load duration class	Order of duration
short-term: C	< 10 hours
medium-term: B	10 hours – 1.5 months
long term: A	> 1.5 months

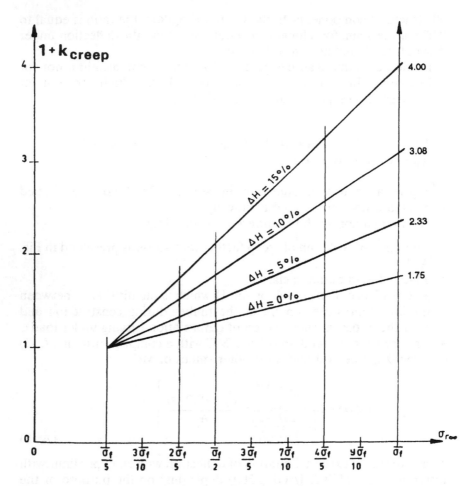

Fig. A.1 Creep factor representation from the French code.

Table A.1 *cont.*

Moisture class	Order of moisture content
indoor, heated: 1	< 12%
under roof: 2	12% – 18%
outdoor: 3	18% – 26%
in water or ground: 4	> 26%

Table A. 2 lists the corresponding creep factors.

Table A. 2 Creep factors, k_{creep} of sawn timber and glued laminated timber for serviceability limit state according to Finnish design code when the reference state is MC I and load class C.

Load duration class	Moisture content class			
	1	2	3	4
short-term: C	0	0	0.3	0.62
medium-term: B	0.3	0.3	0.62	1.17
long term: A	0.62	0.62	1.17	2.71

A.2.3 Sweden – Swedish building code for timber structures, NR1, Chapter 6:4, 1989.

Four load duration classes and four moisture content classes which are very similar to the Finnish case, are considered in Table A.3.

Table A.3

Load duration class	Order of duration
short-term: C	< 10 hours
medium-term: B	10 hours – 1.5 months
long term: A	1.5 months – 10 years
Permanent: P	> 10 years

Moisture class	Relative humidity
0	average < 40%, maximum < 65%
1	average < 65%, maximum < 80%
2	only temporarily higher than 80%
3	other cases

The moisture classes are closely connected to the climatic service conditions of wooden structures.

Table A.4 Creep factors according to Swedish design code.

Load duration class	Moisture content class		
	0 / 1	2	3
short-term: C	0	0.25	0.43
medium-term: B	0.25	0.67	1.0
long term: A	0.67	1.0	1.5
permanent	1.0	1.5	2.33

A2.4 Australia – SAA timber structures code part 1 – design code, Australian Standard 1720.1–1988

The Australian code has different modification factors for stiffnesses (j) associated with major strength properties, as presented. That is the most complete code in dealing with creep. We only detail the case of bending (j_2) as illustrated in table A.6. This code provides continuous creep factors obtained by interpolation from Fig. A.2.

In Table A.5 the moisture class definitions refer only to the initial moisture content of the timber at the beginning of load application.

Table A.5

Load duration class	Order of duration
short-term	< 1 day
medium-term: B	1 day – 1 year
long term: A	> 1 year

Moisture class	Order of moisture content
1	MC at loading < 15%
2	MC at loading 15%–25%
3	MC at loading > 25%

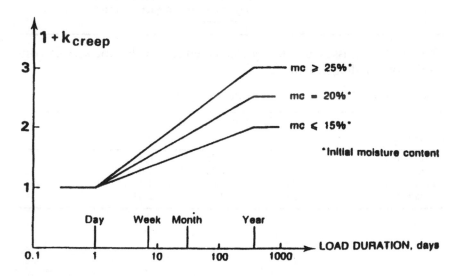

Fig. A.2 Creep factor representation from the Australian code.

Table A.6 Creep factor from the Australian design code

Load duration class	Moisture content class		
	1	2	3
short-term	0	0	0
medium-term	0	interpolation	interpolation
long term	1	interpolation	2

Moreover the Australian code recommends different creep factors for various types of fasteners which are greater than those proposed by Eurocode 5.

A.3 Concluding remarks

From this brief presentation, the following conclusions may be drawn:

1) Only half of the main wood design codes have detailed creep factor tables, closely associated with load duration and moisture content classes.
2) Codes reflect the lack of reliable data for different materials or loading cases, such as:
   • plywood, particle board, fibreboard, waferboard and oriented strand board,
   • shear or perpendicular to the grain behaviour,
   • mechanical joints (glued-in bolts, nail, metal plates,
   • dowels.
3) Codes increase the values of creep if timber is drying under load, but the effects of different or changing temperatures are not included in the moisture classes.

Index

Milton Keynes UK
Ingram Content Group UK Ltd.
UKHW040052071024
449327UK00019B/492